SpringerBriefs in Mathematics

SpringerBriefs present concise summaries of cutting-edge research and practical applications across a wide spectrum of fields. Featuring compact volumes of 50 to 125 pages, the series covers a range of content from professional to academic. Briefs are characterized by fast, global electronic dissemination, standard publishing contracts, standardized manuscript preparation and formatting guidelines, and expedited production schedules.

Typical topics might include:

- A timely report of state-of-the art techniques
- A bridge between new research results, as published in journal articles, and a contextual literature review
- A snapshot of a hot or emerging topic
- An in-depth case study
- A presentation of core concepts that students must understand in order to make independent contributions

SpringerBriefs in Mathematics showcases expositions in all areas of mathematics and applied mathematics. Manuscripts presenting new results or a single new result in a classical field, new field, or an emerging topic, applications, or bridges between new results and already published works, are encouraged. The series is intended for mathematicians and applied mathematicians. All works are peer-reviewed to meet the highest standards of scientific literature.

Titles from this series are indexed by Scopus, Web of Science, Mathematical Reviews, and zbMATH.

Bernardo Uribe Jongbloed ·
Higinio Serrano Garcia ·
Miguel Alejandro Xicoténcatl Merino

Magnetic Equivariant *K*-Theory

Topological Invariants of Magnetic Symmetries

Springer

Bernardo Uribe Jongbloed (iD)
Departamento de Matemáticas y Estadística
Universidad del Norte
Barranquilla, Colombia

Higinio Serrano Garcia (iD)
Institute of Mathematics and Informatics
Bulgarian Academy of Sciences
Sofia, Bulgaria

Miguel Alejandro Xicoténcatl Merino (iD)
Departamento de Matemáticas
CINVESTAV
Mexico City, Mexico

ISSN 2191-8198   ISSN 2191-8201   (electronic)
SpringerBriefs in Mathematics
ISBN 978-3-032-05913-0   ISBN 978-3-032-05914-7   (eBook)
https://doi.org/10.1007/978-3-032-05914-7

This Springer imprint is published by the registered company Springer Nature Switzerland AG
The registered company address is: Gewerbestrasse 11, 6330 Cham, Switzerland

If disposing of this product, please recycle the paper.

Acknowledgments

BU acknowledges the continuous financial support of the Max Planck Institute of Mathematics in Bonn, Germany, the International Center for Theoretical Physics in Trieste, Italy, through its Associates Program, and the Alexander Von Humboldt Foundation in Bonn, Germany.

HS acknowledges the support of SECIHTI of Mexico through the Ph.D. scholarship number CVU: 926934, the support of CONAHCYT in the year 2023 under Frontier Science Grant CF-2023-I-2649., the support of the Simons Foundation, grant SFI-MPS-T-Institutes-00007697, and the Ministry of Education and Science of the Republic of Bulgaria, grant DO1-239/10.12.2024.

All authors acknowledge the support of CONAHCyT of Mexico through grant number CB-2017-2018-A1-S-30345-F-3125.

About This Book

This book is meant to be a gentle introduction to the K-theory groups of complex vector bundles endowed with actions of magnetic groups. It has three main chapters. In Chap. 1 we start with a review of the theory of representations of compact magnetic groups as presented in Wigner's seminal book of group theory and quantum mechanics [1]. We include a review of how to determine the type of an irreducible magnetic representation using the appropriate Schur-Frobenius indicator.

We continue with the bulk of our work in Chap. 2, where we define the magnetic equivariant K-theory groups and we show that the Bott isomorphism, the Thom isomorphism and the degree shift isomorphism, among others, generalize to the magnetic setup. Moreover, we use Atiyah and Segal's approach to the real K-theory groups from the point of view of Clifford modules to determine the coefficients of the magnetic equivariant K-theory as an equivariant cohomology theory. We end this chapter with the description of the Atiyah-Hirzebruch spectral sequence, showcasing the appearance of the Bredon cohomology with coefficients in the Grothendieck group of magnetic representations as its second page.

We finish our work in Chap. 3, where we present an explicit calculation of magnetic equivariant K-theories relevant in the study of two-dimensional altermagnetic materials [2–4]. We give a concise description of the incorporation of the spin-orbit interaction in the setup, and how the magnetic group is enhanced to the appropriate double cover and the K-theory groups become twisted. We focus our attention in the magnetic cyclic group $\mathbb{Z}/4$ and its action on the two-dimensional torus via 4-fold rotations. We explicitly calculate its magnetic equivariant K-theory groups and we show how the K-theory groups are calculated in the case that the spin-orbit interaction is taken into account. The calculation of our last example shows that two-dimensional magnetic crystals which preserve the combination of the 4-fold rotation symmetry composed with time reversal possess a bulk $\mathbb{Z}/2$ invariant that distinguishes trivial insulators from topological ones. This calculation allowed González-Hernández and the first two authors to show that two-dimensional insulating altermagnets with $C_4\mathbb{T}$ symmetry possess a non-trivial phase such as the one of topological insulators [5].

This work assumes a certain degree of mathematical knowledge of representation theory and equivariant algebraic topology. Chapter 1 is based on the theory of complex representations of finite groups [6, 7]. Chapter 2 is based on the algebraic topology of equivariant vector bundles [8, 9], the equivariant cohomology theories (K-theories) they induce [10–14], Clifford modules [15, 16], and the spectral sequences that can be built out of a CW-decomposition of a space [14, 17, 18]. The text has plenty of internal references thus permitting the interested reader to jump ahead in text and allowing the different sections to be read independently.

Here we highlight some of the results that we have collected in this work in relation to the magnetic equivariant K-theory.

Classification of Magnetic Representations

We give a summary of Wigner's account [1] on the classification of magnetic representations in Theorem 1.2. For an irreducible magnetic representation V of the magnetic group (G, ϕ), the vector space

$$\mathrm{End}_{\mathbf{Rep}(G,\phi)}(V) \cong \mathbb{F}, \tag{1}$$

where $\mathbb{F} \in \{\mathbb{R}, \mathbb{C}, \mathbb{H}\}$ determines the type. The Schur-Frobenius indicator of Definition 1.5 gives a formula to determine its type.

Coefficients in Magnetic Equivariant K-Theory

The magnetic equivariant K-theory of a point splits into real, complex, and symplectic K-theories according to the number of irreducible representations of the magnetic group of real, complex and quaternionic type. The formula appears in Eq. 2.176 and reads:

$$\mathbf{K}_G^*(pt) \cong KO^*(pt)^{\oplus n_{\mathbb{R}}} \oplus KU^*(pt)^{\oplus n_{\mathbb{C}}} \oplus KSp^*(pt)^{\oplus n_{\mathbb{H}}} \tag{2}$$

where $n_{\mathbb{F}}$ is the number of irreducible representations of $\mathbb{F}$-type. The proof is on Theorem 2.5.

Degree Shift Isomorphism and 8-Periodicty

The isomorphism between real K-theory and symplectic K-theory, $K\mathbb{R}^* \cong KSp^{*+4}$, is generalized to the context of magnetic equivariant K-theory, inducing an isomorphism of degree 4

$$\mathbf{K}_G^*(X) \xrightarrow{\ \cong\ } \overset{\widehat{G}}{\mathbf{K}_G^{*+1}(X)}.$$ (3)

where the right hand side is the $\widehat{G}$-twisted magnetic G-equivariant K-theory with $\widehat{G}$ being defined as the pullback group $\phi^*(\mathbb{Z}/4)$. Whenever the degree shift isomorphism is applied twice, the 8-periodicity of the magnetic equivariant K-theory groups is obtained

$$\mathbf{K}_G^*(X) \xrightarrow{\ \cong\ } \mathbf{K}_G^{*+8}(X).$$ (4)

The proof appears in Theorem 2.6 and it is based on the degree shift isomorphism at the level of Clifford modules of Theorem 2.7. Whenever the $\mathbb{Z}/2$-extension $\widehat{G}$ is trivial, then the magnetic equivariant K-theory becomes 4-periodic. This is Prop. 2.6.

Thom Isomorphism

For a magnetic (G, ϕ)-equivariant vector bundle $E \xrightarrow{\ p\ } X$ over a compact G-space X, the homomorphism

$$\varphi_* : \mathbf{K}_G^{-p}(X) \xrightarrow{\ \cong\ } \mathbf{K}_G^{-p}(E)$$
$$F \mapsto \Lambda_E^* \otimes p^* F$$ (5)

is the Thom isomorphism. Here Λ_E^* is the Thom class of E and the proof appears in Theorem 2.9.

Atiyah-Hirzebruch Spectral Sequence

Calculations are important in condensed matter physics and therefore a procedure for determining the magnetic equivariant K-theory is relevant. The Atiyah-Hirzebruch spectral sequence is one tool that permits to calculate such groups. Filtering $\mathbf{K}_G^*(X)$ using the G-CW decomposition of X, we obtain a spectral sequence whose second page is

$$E_2^{n,t} \cong H^n\big(X; \mathbf{K}_G^t\big) \implies \mathbf{K}_G^*(X),$$ (6)

and that abuts to $\mathbf{K}_G^*(X)$. The second page is Bredon's equivariant cohomology with coefficients in the group of representations of the magnetic group. The proof appears in Theorem 2.13.

$C_4\mathbb{T}$ Symmetry in 2-Dimensional Torus

The symmetry C_4 stands for a 4-fold rotation on the 2-dimensional torus, $\mathbb{T}$ stands for time reversal symmetry and $C_4\mathbb{T}$ is the composition. We calculate the magnetic $\langle C_4\mathbb{T}\rangle$-equivariant K-theory of the 2-dimensional torus T^2 in both the context of no spin orbit interaction in Prop. 3.1 and spin orbit interaction in Prop. 3.2:

n	0	-1	-2	-3
$\mathbf{K}^n_{\mathbb{Z}/4}(T^2)$	$\mathbb{Z}^4$	$\mathbb{Z}/2$	$(\mathbb{Z}\oplus\mathbb{Z}/2)\oplus\mathbb{Z}$	0
$^{\mathbb{Z}/8}\mathbf{K}^n_{\mathbb{Z}/4}(T^2)$	$\mathbb{Z}^2\oplus\mathbb{Z}/2$	0	$\mathbb{Z}^3\oplus\mathbb{Z}$	$\mathbb{Z}/2$

$$(7)$$

In both cases, the magnetic equivariant K-theory groups are 4-periodic. Here we highlight that the bulk invariants for $n = 0$ are trivial for the first case and $\mathbb{Z}/2$ for second case, while for $n = -2$ the bulk invariants are $\mathbb{Z}$ in both cases. Bulk invariants are the ones that are due to the presence of the 2-cell.

Introduction

Symmetry in physics has long been a cornerstone for understanding the laws of nature, providing a profound link between invariance and conservation laws, as elegantly formalized in Emmy Noether's theorem. This fundamental result establishes that every differentiable symmetry of the action of a physical system corresponds to a conserved quantity, such as energy, momentum, or angular momentum. These principles not only underpin classical and quantum mechanics but also extend to more intricate systems, where symmetry governs complex interactions and emergent phenomena.

Magnetic symmetries play a critical role in a wide array of physical systems, from the electronic properties of materials to the classification of topological phases of matter. These symmetries, often described through magnetic spaces groups, extend the familiar framework of crystallographic symmetry by incorporating time-reversal operations. The study of representations of magnetic groups provides a foundational understanding of how such symmetries manifest in physical systems, particularly in contexts like electronic band structures and spin configurations.

Equally significant is the role of vector bundles in mathematics and physics. When enriched with symmetries, as in the case of magnetic equivariant vector bundles, these structures become central to topological approaches in condensed matter physics, including the classification of topological insulators and superconductors. The application of K-theory, a tool from algebraic topology designed to understand topological invariants of bundles, allows for a systematic study of electronic band bundles (Bloch bundles), leading to profound insights into their properties and invariants.

In the quest to understand the phases of matter, traditional classifications relied on concepts such as symmetry breaking. However, in the recent decades, a new paradigm has emerged, based on topological features that remain unchanged under continuous deformations. These properties have reshaped our understanding of matter in profound ways.

Topological invariants, mathematical quantities that remain constant under such transformations, have proven crucial in distinguishing different phases of matter beyond conventional classifications.

The journey toward recognizing the role of topology in condensed matter physics began with theoretical predictions in the 1980s [19–21], particularly in the study of the quantum Hall effect. In this phenomenon, the electrical conductivity of a two-dimensional electron system exhibits quantized plateaus, a behavior explained by a topological invariant known as the Chern number of the material's vector bundle of valence states, or Bloch bundle.

Further explorations [22, 23] revealed a broader class of materials, including topological insulators, characterized by conducting surfaces despite being insulating in their interiors. This behavior defied previous theoretical expectations and highlighted the power of topological invariants in predicting physical properties. Here, the invariant which characterizes a state as trivial or non trivial band insulator is a number in the group $\mathbb{Z}/2$ (regardless of whether the state exhibits or does not exhibit a quantum spin Hall effect). This number is not zero whenever the valence bands generate the $\mathbb{Z}/2$-invariant, which was shown to live in Atiyah's real K-theory of the three-dimensional torus [24]. This invariant is known as the Kane-Mele invariant.

The theoretical and experimental advancements were highlighted in the 2016 Nobel Prize in Physics, awarded to David J. Thouless, F. Duncan M. Haldane, and J. Michael Kosterlitz [25]. Their pioneering work on topological phases of matter and phase transitions unveiled new quantum states governed by topological principles. This recognition underscored the profound interplay between mathematics and physics, showcasing how abstract mathematical concepts can illuminate the structure of the physical universe.

The study of topological invariants of crystals, whether magnetic or not, relies on an explicit understanding of the symmetry group of the crystal and the equivariant K-theory groups of the two-dimensional and three-dimensional torus. The relevant symmetry groups include the spatial symmetries of the crystal as well as symmetries that, when composed with the time-reversal operator, preserve the Hamiltonian. In the physics literature, these groups are known as magnetic groups or Shubnikov groups [26], and in some mathematics literature as orientifold groups [27]. Adopting the physics term "magnetic groups" for this groups, we define their associated equivariant K-theory groups which we will refer to as "magnetic equivariant K-theory groups" to distinguish them from the classical complex equivariant K-theory groups [11].

The development of magnetic equivariant K-theory can be traced back to Atiyah's seminal 1966 article on K-theory and reality [28]. In this foundational work, Atiyah studied the smallest magnetic point group (the cyclic group with two elements) and complex vector bundles with such symmetry, known as vector bundles with involution. One year later, Atiyah and Segal [12] extended this framework to include magnetic groups formed by the semidirect product of a group with $\mathbb{Z}/2$, where the first factor acts linearly and the second antilinearly on the complex vector bundles. This generalization allowed for the inclusion of both spatial and time-reversal symmetries in vector bundles. Subsequently, in 1970, Karoubi [29] further generalized the theory to encompass any magnetic point group.

In the 1980s and 1990s, researchers began using additional mathematical tools, such as noncommutative geometry and the K-theory of C^*-algebras (algebraic K-theory), to study condensed matter systems. These approaches offered new insights and methodologies, as seen in works like [30, 31] and the references therein, highlighting the growing intersection of mathematical physics and topology.

In 2013, Freed and Moore [32] applied magnetic equivariant K-theory to the study of topological phases of matter, demonstrating its significance in describing physical systems with symmetries. Building on this foundation, Gomi [33] expanded upon the K-theory framework introduced by Freed and Moore, developing it in greater detail. He introduced general twistings on the groups as cocycles and within the structure of vector bundles, extending the theory to encompass a wider range of physical phenomena. These advancements further enriched the applicability and mathematical depth of magnetic equivariant K-theory, solidifying its role in both mathematics and physics.

Our main objective in this work is to showcase the properties of the magnetic equivariant K-theory in a simple, clear and concise way so that it can be used as a reference for non-experts in algebraic topology. In particular, we had in mind the condensed matter physicist interested in determining the topological invariants of prescribed magnetic crystals. We have thus avoided the use of categorical language such as that of groupoids done in [33] and have focused our attention on highlighting the most interesting features of the magnetic equivariant K-theory.

We decided to write our work in a style as similar as possible to that of Atiyah and Segal in their papers [28, 11, 12]. In some sense, this work simply puts Atiyah and Segal's work in the framework of magnetic groups and shows that all of their results in real and equivariant K-theory generalize to the magnetic equivariant case. We put great care in highlighting the simplicity and clarity of the theory so that our work can be used as a reference manual for calculational purposes.

Contents

Abbreviations

In this work several mathematical structures have been defined. In order to make the text easily accessible, we list the most important symbols that we have used throughout our work. They will come with a brief explanation and a cross-reference to where they are introduced for the first time. We highlight that *symbols in boldface are the ones associated to magnetic information.*

(G, ϕ)	Magnetic group. Consists of a group G together with a surjective homomorphism $\phi : G \to \mathbb{Z}/2$. Definition 1.1.
G_0	Core of the magnetic group. Kernel of $\phi : G \to \mathbb{Z}/2$ consisting of the elements that act unitarily on magnetic representations. Definition 1.1.
$\mathbf{Rep}(G, \phi)$	Category of magnetic representations of (G, ϕ). Definition 1.3.
$\mathrm{Rep}(G_0)$	Category of complex representations of the group G_0. Definition 1.3.
$\mathbb{K}$	Complex conjugation operator.
$\mathbf{R}(G, \phi)$	Grothendieck group of isomorphism classes of magnetic representations of (G, ϕ), also denoted $\mathbf{R}(G)$. Definition 1.3, Eq. (1.72).
$\mathbf{R}(G, \phi; \mathbb{F})$	Grothendieck group of isomorphism classes of magnetic representations of (G, ϕ) of type $\mathbb{F}$ for $\mathbb{F} \in \{\mathbb{R}, \mathbb{C}, \mathbb{H}\}$, also denoted $\mathbf{R}(G, \mathbb{F})$. Eq. (1.72).
$R(G_0)$	Grothendieck ring of complex representations of the group G_0. Definition 1.3.
$\mathbb{R}^{p,q}$	Real vector space $\mathbb{R}^{q+p} = \mathbb{R}^q \oplus i\mathbb{R}^p$ with the involution mapping $x + iy$ to $x - iy$.
$D^{p,q}$	Unit disk in $\mathbb{R}^{p,q}$.
$S^{p,q}$	Unit sphere in $\mathbb{R}^{p,q}$.
X	G-CW-complex.
$\mathbf{Vect}_{(G,\phi)}(X)$	Category of magnetic (G, ϕ)-equivariant complex vector bundles. Also denoted $\mathbf{Vect}_G(X)$. Definition 2.3.
$\mathbf{K}_G(X)$	Magnetic G-equivariant K-theory of X. Definition 2.3.
$KU_{G_0}(X)$	G_0-equivariant complex K-theory of X [11]
$K\mathbb{R}^*$	Atiyah's real K-theory of spaces with involution [28].

$K\mathbb{H}^*$	Quaternionic K-theory of spaces with involution.
KO^*	K-theory of real vector bundles.
KS_{p*}	Dupont's symplectic K-theory [34].
$\widetilde{\mathbf{K}}_G X$	Reduced magnetic equivariant K-theory. Prop. 2.3.
$C_{p,q}$	Clifford algebra over $\mathbb{R}^{p+q}$ with a negative quadratic form for q variables and positive for p variables. Eq. (2.128).
$\widehat{M}^{p,q}(G,\phi)$	Grothendieck group of magnetic graded $C_{p,q}[G,\phi]$-modules. Definition 2.9.
$(\widetilde{G},\widetilde{\phi})$	Magnetic central extension of (G,ϕ) by $A=(\mathbb{Z}/2)^k$. See Sect. 2.8.
$^{(\widetilde{G},\chi)}\mathbf{K}_G^*(X)$	$(\widetilde{G},\chi)$-twisted magnetic G-equivariant K-theory groups of X with χ a fixed character of A. Definition 2.11.
$(\widehat{G},\widehat{\phi})$	$\mathbb{Z}/2$-central magnetic extension of (G,ϕ) defined as the pullback $\phi^*(\mathbb{Z}/4)$. Eq. (2.186).
$^{\widehat{G}}\mathbf{K}_G^*(X)$	Twisted magnetic equivariant K-theory which is the image of the degree shift isomorphism. Theorem 2.6.
$^{\widehat{G}}\widehat{M}_{\mathbb{R}}^{p,q}(G,\phi)$	Grothendieck group of magnetic graded $(\widehat{G},\widehat{\sigma})$-twisted $C_{p,q}[G,\phi]$-modules. Definition 2.12.
Λ_E^*	Thom class of the bundle $E \to X$; it defines an element in $\mathbf{K}_G(E)$. Example 2.10.
$H^*(X;\mathbf{K}_G^*)$	Bredon's equivariant cohomology with coefficients in magnetic equivariant K-theory. Theorem 2.13.
$\widetilde{G}^{soc}$	$\mathbb{Z}/2$ central extension of G defined by the spin orbit interaction. Definition 3.1.
$^{\widetilde{G}^{soc}}\mathbf{K}_G^*((S^1)^3)$	Twisted magnetic equivariant K-theory groups for the 3-dimensional torus in the presence of spin orbit interaction. Definition 3.2.

References

1. Eugene P. Wigner. Group theory and its application to the quantum mechanics of atomic spectra, volume Vol. 5 of Pure and Applied Physics. Academic Press, New York-London, 1959. Expanded and improved ed. Translated from the German by J. J. Griffin.
2. Libor Šmejkal, Jairo Sinova, and Tomas Jungwirth. Beyond conventional ferromagnetism and antiferromagnetism: A phase with nonrelativistic spin and crystal rotation symmetry. Phys. Rev. X, 12:031042, Sep 2022.
3. Rafael González-Hernández, Libor Šmejkal, Karel Výborný, Yuta Yahagi, Jairo Sinova, Tomáš Jungwirth, and Jakub Železný. Efficient electrical spin splitter based on nonrelativistic collinear antiferromagnetism. Phys. Rev. Lett., 126:127701, Mar 2021.
4. Igor Mazin. Editorial: Altermagnetism—a new punch line of fundamental magnetism. Phys. Rev. X, 12:040002, Dec 2022.
5. Rafael González-Hernández, Higinio Serrano, and Bernardo Uribe. Spin chern number in altermagnets. Phys. Rev. B, 111:085127, Feb 2025.
6. Jean-Pierre Serre. Linear representations of finite groups, volume Vol. 42 of Graduate Texts in Mathematics. Springer-Verlag, New York-Heidelberg, french edition, 1977.
7. William Fulton and Joe Harris. Representation theory, volume 129 of Graduate Texts in Mathematics. Springer-Verlag, New York, 1991. A first course, Readings in Mathematics.

8.   Dale Husemöller. Fibre bundles, volume 20 of Graduate Texts in Mathematics. SpringerVerlag, New York, third edition, 1994.

9.   Wolfgang Lück and Bernardo Uribe. Equivariant principal bundles and their classifying spaces. Algebr. Geom. Topol., 14(4):1925–1995, 2014.

10. M. F. Atiyah. K-theory. W. A. Benjamin, Inc., New York-Amsterdam, 1967. Lecture notes by D. W. Anderson.

11. Graeme Segal. Equivariant K-theory. Inst. Hautes Études Sci. Publ. Math., (34):129–151, 1968.

12. M. F. Atiyah and G. B. Segal. Equivariant K-theory and completion. J. Differential Geometry, 3:1–18, 1969.

13. Glen E. Bredon. Equivariant cohomology theories, volume No. 34 of Lecture Notes in Mathematics. Springer-Verlag, Berlin-New York, 1967.

14. Dieter Degrijse, Markus Hausmann, Wolfgang Luck, Irakli Patchkoria, and Stefan Schwede. Proper equivariant stable homotopy theory. Mem. Amer. Math. Soc., 288(1432):vi+142, 2023.

15. M. F. Atiyah, R. Bott, and A. Shapiro. Clifford modules. Topology, 3(suppl):3–38, 1964.

16. H. Blaine Lawson, Jr. and Marie-Louise Michelsohn. Spin geometry, volume 38 of Princeton Mathematical Series. Princeton University Press, Princeton, NJ, 1989.

17. Graeme Segal. Categories and cohomology theories. Topology, 13:293–312, 1974.

18. D. Husemöller, M. Joachim, B. Jurčo, and M. Schottenloher. Basic bundle theory and Kcohomology invariants, volume 726 of Lecture Notes in Physics. Springer, Berlin, 2008. With contributions by Siegfried Echterhoff, Stefan Fredenhagen and Bernhard Krötz.

19. D. J. Thouless, M. Kohmoto, M. P. Nightingale, and M. den Nijs. Quantized hall conductance in a two-dimensional periodic potential. Phys. Rev. Lett., 49:405–408, Aug 1982.

20   F. D. M. Haldane. Model for a quantum hall effect without landau levels: Condensed-matter realization of the "parity anomaly"". Phys. Rev. Lett., 61:2015–2018, Oct 1988.

21. Cui-Zu Chang, Jinsong Zhang, Xiao Feng, Jie Shen, Zuocheng Zhang, Minghua Guo, Kang Li, Yunbo Ou, Pang Wei, Li-Li Wang, Zhong-Qing Ji, Yang Feng, Shuaihua Ji, Xi Chen, Jinfeng Jia, Xi Dai, Zhong Fang, Shou-Cheng Zhang, Ke He, Yayu Wang, Li Lu, Xu-Cun Ma, and Qi-Kun Xue. Experimental observation of the quantum anomalous hall effect in a magnetic topological insulator. Science, 340(6129):167–170, April 2013.

22. C. L. Kane and E. J. Mele. Quantum spin hall effect in graphene. Phys. Rev. Lett., 95:226801, Nov 2005.

23. Elizabeth Gibney and Davide Castelvecchi. Physics of 2d exotic matter wins nobel. Nature, 538(7623):18, October 2016.

24. B. Andrei Bernevig and Shou-Cheng Zhang. Quantum spin hall effect. Phys. Rev. Lett., 96:106802, Mar 2006.

25. C. L. Kane and E. J. Mele. Z2 topological order and the quantum spin hall effect. Phys. Rev. Lett., 95:146802, Sep 2005.

26. A.V. Shubnikov, N.V. Belov, W.T. Holser, and Institut kristallografii im. A.V. Shubnikova. Colored Symmetry. Macmillan, 1964.

27. Krzysztof Gawⸯedzki, Rafał R. Suszek, and Konrad Waldorf. Bundle gerbes for orientifold sigma models. Adv. Theor. Math. Phys., 15(3):621–687, 2011.

28. M. F. Atiyah. K-theory and reality. Quart. J. Math. Oxford Ser. (2), 17:367–386, 1966.

29. Max Karoubi. Sur la K-théorie équivariante. In Séminaire Heidelberg-Saarbrücken-Strasbourg sur la K-Théorie, pages 187–253, Berlin, Heidelberg, 1970. Springer Berlin Heidelberg.

30. Jean Bellissard. K-theory of C ∗-algebras in solid state physics. In Statistical mechanics and field theory: mathematical aspects (Groningen, 1985), volume 257 of Lecture Notes in Phys., pages 99–156. Springer, Berlin, 1986.

31. J. Bellissard, A. van Elst, and H. Schulz-Baldes. The noncommutative geometry of the quantum Hall effect. J. Math. Phys., 35(10):5373–5451, 1994. Topology and physics.

32. Daniel S. Freed and Gregory W. Moore. Twisted equivariant matter. Ann. Henri Poincaré, 14(8):1927–2023, 2013.

33. Kiyonori Gomi. Freed Moore K theory. Comm. Anal. Geom., 31(4):979–1067, 2023

34. Johan L. Dupont. Symplectic bundles and KR-theory. Math. Scand., 24:27–30, 1969

Chapter 1
Magnetic Groups and Their Representations

Abstract In this chapter we define magnetic groups and their finite dimensional magnetic representations. We describe the morphisms between magnetic representations and we show the classification of irreducible magnetic representations by their real, complex or quaternionic type.

Keywords Magnetic group · Magnetic representation · Corepresentation

In this chapter we develop the theory of representations of magnetic point groups, namely groups that come equipped with a surjective homomorphism onto $\mathbb{Z}/2$. These groups emerge as symmetries in quantum systems, particularly in contexts involving time-reversal symmetry or combined spatial and time-reversal symmetries [1]. Magnetic point groups act via complex linear or antilinear maps, depending on whether a group element lies in the kernel of the homomorphism. This introduces an added layer of complexity in their representation theory.

A key difference between the representations of magnetic point groups and ordinary groups lies in the structure of the morphisms between irreducible representations, as explained by Schur's lemma. For non magnetic (ordinary) groups, the set of homomorphisms of an irreducible representation consists solely of complex multiples of the identity matrix. However, in the case of magnetic groups, the set of morphisms of irreducible representations can take one of three forms: real, complex, or quaternion multiples of the identity.

This distinction is not merely formal; it has deep physical implications. In quantum systems, the presence of time-reversal symmetry or combined symmetries can lead to phenomena like Kramers' degeneracy, where pairs of states remain degenerate due to time-reversal symmetry, even in the absence of other symmetries [2]. The analysis of how representations change under the action of antilinear maps is crucial for understanding such phenomena, particularly in systems with strong spin-orbit interactions or in materials that exhibit topological phases [3]. Thus, the study of morphisms in this more general setting of magnetic point groups opens a pathway to understanding physical systems that cannot be described by ordinary group theory alone.

1.1 Magnetic Groups

Definition 1.1 A **magnetic group** consists of a pair (G, ϕ_G) where G is a group and $\phi_G : G \longrightarrow \mathbb{Z}/2$ a surjective homomorphism. The group $G_0 := \ker \phi_G$ will be called the **core group** of the magnetic group. This information is usually depicted in the short exact sequence of groups

$$1 \longrightarrow G_0 \longrightarrow G \overset{\phi_G}{\longrightarrow} \mathbb{Z}/2 \longrightarrow 0. \tag{1.1}$$

A **morphism** between the magnetic groups (G, ϕ_G) and (H, ϕ_H) is a homomorphism $f : G \longrightarrow H$ such that $\phi_H \circ f = \phi_G$.

We will denote by **MGrp** the category of magnetic groups and their morphisms.

A **subgroup** of a magnetic group (G, ϕ_G) is a magnetic group (H, ϕ_H) such that $H \leq G$ and this inclusion is a morphism in **MGrp**. It is a **normal subgroup** if $H \trianglelefteq G$.

Note that not every subgroup of H_0 of G_0 comes from a magnetic subgroup (H, ϕ). For example, take the magnetic group $1 \longrightarrow \mathbb{Z}/4 \longrightarrow \mathbb{Z}/8 \longrightarrow \mathbb{Z}/2 \longrightarrow 1$ and the subgroup $\mathbb{Z}/2$ of $\mathbb{Z}/4$. This group $\mathbb{Z}/2$ is not the core of any magnetic subgroup of $(\mathbb{Z}/8, \phi_{\mathbb{Z}/8})$.

In condensed matter physics, the magnetic groups that classify the symmetries of a crystal which take into account additional symmetries such as time-reversal, are called **magnetic space groups**. Whenever the translational symmetries are mod-out, the quotient group is called the **magnetic point group** (see [4, sect. 7]).

Example 1.1 Perhaps the most well-known example of a magnetic group is the **magnetic general linear group**. For a complex vector space V denote by $\mathrm{MGL}(V)$ the group of invertible linear or antilinear operators on V where $\phi : \mathrm{MGL}(V) \to \mathbb{Z}/2$ maps an operator to 0 if it is complex linear and 1 if it is antilinear. Its core group $\mathrm{MGL}(V)_0 = \mathrm{GL}(V)$ is the group of invertible linear operators of V.

The general linear group with the complex conjugation automorphism is an example of what is known as a *Real group* [5]. That is, a group G_0 with an automorphism $\tau : G_0 \longrightarrow G_0$ such that $\tau^2 = \mathrm{id}$. The group $G_0 \rtimes \mathbb{Z}/2$ together with the projection π_2 to $\mathbb{Z}/2$ becomes a magnetic group.

1.2 Representations of Magnetic Groups

The representation theory of magnetic groups extends the classical representation theory of ordinary groups by incorporating time-reversal symmetry. Magnetic groups arise in systems where time-reversal symmetry or a combination of spatial and time-reversal symmetries plays a significant role, such as in quantum systems with magnetic or spin-orbit coupling effects.

Definition 1.2 A **representation** of a magnetic group (G, ϕ) on a finite-dimensional complex vector space V is a morphism of magnetic groups $\rho : G \longrightarrow MGL(V)$.

We also call ρ a *magnetic representation*.

Example 1.2 Consider the magnetic group $(G_0 \times \mathbb{Z}/2, \pi_2)$ where $\phi = \pi_2$ is the projection onto the second factor and the core group is G_0.

Take any representation $\rho : G_0 \times \mathbb{Z}/2 \to MGL(V)$ and note that $D(g) := \rho(g, 0)$ for $g \in G_0$ defines a complex representation of the group G_0. Now set $J := \rho(e, 1)$, which is an antilinear operator commuting with D such that $J^2 = \mathrm{id}_V$. So $D : G_0 \longrightarrow GL(V)$ is a complex representation and $J : V \longrightarrow V$ an anti-linear involution which commutes with D. Therefore, the pair (D, J) is a **real representation** of G_0 in the sense of Atiyah [6].

Clearly, we have the bijection

$$\left\{ \begin{array}{c} \text{Real representations} \\ \text{of the group } G_0 \end{array} \right\} \longleftrightarrow \left\{ \begin{array}{c} \text{Representations of the} \\ \text{magnetic point group } (G_0 \times \mathbb{Z}/2, \pi_2) \end{array} \right\}. \quad (1.2)$$

Remark 1.1 Let $(D : G_0 \longrightarrow GL(V), J : V \longrightarrow V)$ be a real representation of G_0. Then $W = \{v \in V \mid J(v) = v\}$ is a vector space over the real numbers, which is closed under the action of D. Thus, G_0 acts on W by matrices with real coefficients. On the other hand, if W is a vector space, over the real numbers and $D : G_0 \longrightarrow GL(W)$ is a representation over the real numbers, then $(D \otimes 1 : G_0 \longrightarrow GL(W \otimes \mathbb{C}), 1 \otimes \mathbf{K} : W \otimes \mathbb{C} \longrightarrow W \otimes \mathbb{C})$ is a real representation of G_0. These constructions are inverses of each other.

Example 1.3 Take the magnetic point group $(\mathbb{Z}/4, \mathrm{mod}\, 2)$ and note that there are two representations that come straight to mind. One can have the one-dimensional representation where the generator of $\mathbb{Z}/4$ is mapped to $\mathbb{K}$, the complex conjugation operator in complex vector spaces, or the two-dimensional representations where the generator is mapped to $\begin{pmatrix} 0 & 1 \\ -1 & 0 \end{pmatrix} \cdot \mathbb{K}$.

1.3 Morphisms of Representations

Here we give the notion of a morphism of representations of magnetic groups and that of an irreducible one. Next we state and prove an appropriate version of Schur's lemma in this context. The material presented here is based on the pioneering work of Wigner [7] who called representations of magnetic groups **corepresentations**.

Definition 1.3 A **morphism** of representations $\rho_i : G \longrightarrow MGL(V_i)$ $i = 1, 2$ of the magnetic group (G, ϕ) is a (complex) linear transformation $T : V_1 \longrightarrow V_2$ such that $T\rho_1(g) = \rho_2(g)T$ for all $g \in G$.

Two representations ρ_1, ρ_2 are **isomorphic** if there exists a morphism of representations which is invertible.

For (G, ϕ) a fixed magnetic group denote:

- **Rep**(G, ϕ) the category of representations of magnetic groups with their morphisms. Denote its Grothendieck group of isomorphism classes of representations by **R**(G, ϕ).
- $\mathrm{Hom}_{\mathbf{Rep}(G,\phi)}(V, W)$ the $\mathbb{R}$-vector space of morphisms of representations and by $\mathrm{End}_{\mathbf{Rep}(G,\phi)}(V)$ the endomorphisms of a representation V.
- $\mathrm{Rep}(G_0)$ the category of representations of the core group G_0 and by $\mathrm{R}(G_0)$ its Grothendieck ring of complex representations.

Example 1.4 Let $\rho : G \longrightarrow \mathrm{MGL}(V)$ be a representation of the magnetic group (G, ϕ) and $z \in \mathbb{C}$ a complex number such that $|z| = 1$. Define the map

$$\rho_1(g) := \overline{z} \cdot \rho(g) \cdot z = \overline{z} \cdot \mathbb{K}^{\phi(g)}(z) \cdot \rho(g) = \begin{cases} \rho(g) & \text{if } \phi(g) = 0 \\ \overline{z}^2 \rho(g) & \text{if } \phi(g) = 1 \end{cases} \qquad (1.3)$$

that is, ρ_1 and ρ differ only by a factor of $\overline{z}^2$ in the antilinear part.

This map is a representation of the magnetic point group because

$$\rho_1(gh) = \overline{z} \cdot \rho(gh) \cdot z \qquad (1.4)$$
$$= \overline{z} \cdot \rho(g) \cdot \rho(h) \cdot z \qquad (1.5)$$
$$= \overline{z} \cdot \rho(g) \cdot z\overline{z} \cdot \rho(h) \cdot z \qquad (1.6)$$
$$= \rho_1(g) \cdot \rho_1(h). \qquad (1.7)$$

Now, if T denotes multiplication by z then

$$T\rho_1(g) = z\overline{z}\rho(g)z = \rho(g)z = \rho(g)T, \qquad (1.8)$$

and therefore the representations ρ and ρ_1 are isomorphic. Hence the antilinear part of representations of magnetic groups can be multiplied by any complex number of norm one and the isomorphism class of the representation remains unaffected.

If we have a representation V of the magnetic point group (G, ϕ), then the **restriction** to the core group G_0 is a representation of G_0 and is denoted by $\mathrm{res}(V)$. This restriction map defines a functor at the level of the categories of representations.

$$\mathbf{Rep}(G, \phi) \xrightarrow{\ \mathrm{res}\ } \mathrm{Rep}(G_0). \qquad (1.9)$$

A **subrepresentation** of a representation V of the magnetic point group (G, ϕ) is a subspace $W \subset V$ such that the restriction of the G-action to W is also a representation of (G, ϕ). A representation V of (G, ϕ) is **irreducible** if its only representations are V and 0.

Now, the category of representations of finite magnetic groups is semisimple.

Proposition 1.1 *Every representation of a finite magnetic group is a sum of irreducible representations.*

Which follows from Maschke's lemma applied to representations of magnetic groups.

Lemma 1.1 (Maschke) *If W is a subrepresentation of V of a finite magnetic group, then there exists a subrepresentation W' of V such that $V \cong W \oplus W'$ as representtions of (G, ϕ).*

Proof Write V as a direct sum $V = W \oplus U$ (with U not necessarily a subrepresentation) and consider the projection $\pi_1 : V \longrightarrow W$ given by $\pi_1(w + u) = w$. Define

$$\pi : V \longrightarrow W$$

$$v \longmapsto \sum_{g \in G} g \cdot \pi_1(g^{-1} \cdot v). \tag{1.10}$$

The morphism π is in $\mathrm{Hom}_{\mathbf{Rep}(G,\phi)}(V, W)$ because if $h \in G$ and $v \in V$, then

$$\pi(h \cdot v) = \sum_{g \in G} g \cdot \pi_1(g^{-1} \cdot (h \cdot v)) \tag{1.11}$$

$$= \sum_{g \in G} g \cdot \pi_1((g^{-1}h) \cdot v) \tag{1.12}$$

$$= \sum_{g \in G} (hh^{-1}g) \cdot \pi_1((h^{-1}g)^{-1} \cdot v) \tag{1.13}$$

$$= h \cdot \sum_{g \in G} (h^{-1}g) \cdot \pi_1((h^{-1}g)^{-1} \cdot v) \tag{1.14}$$

$$= h \cdot \sum_{l \in G} l \cdot \pi_1(l^{-1} \cdot v) \tag{1.15}$$

$$= h \cdot \pi(v) \tag{1.16}$$

The morphism π is surjective; this follows since $\pi_1(w) = w$ for $w \in W$, so $\pi(w) = |G|w$. If we denote $W' = \ker \pi$, then we have the decomposition as a direct sum of representations of (G, ϕ)

$$V \cong \mathrm{Im}\,\pi \oplus \ker \pi = W \oplus W'. \tag{1.17}$$

$\sqcup$

One key step required to classify the irreducible representations of magnetic groups (G, ϕ) is to construct the induced representations of the core $G_0 \leq G$.

Definition 1.4 Let (G, ϕ) be a finite magnetic point group and V a representation of the core group G_0. The **induced representation** of V is the complex vector space

$$\mathrm{Ind}_{G_0}^{(G,\phi)} V := \mathbb{C}[G, \phi] \otimes_{\mathbb{C}[G_0]} V \tag{1.18}$$

where $\mathbb{C}[G, \phi]$ is the **regular representation** of (G, ϕ), i.e. the complex vector space generated by G together the action of (G, ϕ) given by

$$g \cdot \left(\sum_i z_i g_i \right) = \sum_i \mathbf{K}^{\phi(g)}(z_i) g g_i, \tag{1.19}$$

and whose action of (G, ϕ) on $\mathrm{Ind}_H^{(G,\phi)} V$ is given by its action on the first factor.

For a fixed element $a_0 \in G \backslash G_0$, we have the following presentation for the induced representation

$$\mathrm{Ind}_{G_0}^{(G,\phi)} V = V \oplus a_0 V \tag{1.20}$$

where $a_0 V = \{a_0 v \mid v \in V\}$ is a complex vector space with the scalar product given by

$$z \cdot a_0 v = a_0 \bar{z} v. \tag{1.21}$$

The vector space $a_0 V$ is a representation of G_0 with action

$$g \cdot a_0 v = a_0 (a_0^{-1} g a_0 v), \tag{1.22}$$

and this representation is called the **conjugate representation** of V by a_0.

If $\rho : G_0 \longrightarrow \mathrm{GL}(V)$ is the representation of the core, then the conjugate representation $a_0 V$ can be written by the homomorphism $\rho' : G_0 \longrightarrow \mathrm{GL}(V)$ where

$$\rho'(g) = \overline{\rho(a_0^{-1} g a_0)}. \tag{1.23}$$

Here we have implicitly assumed the isomorphism $a_0 V \cong V$, $a_0 v \mapsto v$.

The action of (G, ϕ) on $V \oplus a_0 V$ is given as follows

1. If $g \in G_0$, then $g \cdot (v, a_0 w) = \big(gv, a_0(a_0^{-1} g a_0 w)\big)$. We can write this formula as

$$g \cdot \begin{pmatrix} v \\ w \end{pmatrix} = \begin{pmatrix} \rho(g) & 0 \\ 0 & \rho'(g) \end{pmatrix} \begin{pmatrix} v \\ w \end{pmatrix}. \tag{1.24}$$

2. If $g \notin G_0$, then we can write $g = a_0 g'$ for some $g' \in G_0$ and

$$g \cdot (v, a_0 w) = a_0 g' \cdot (v, a_0 w)$$
$$= a_0 \cdot \left(g'v, a_0(a_0^{-1} g' a_0 w) \right)$$
$$= (a_0 g' a_0 w, a_0 g' v)$$
$$= \left(g a_0 w, a_0(a_0^{-1} g v) \right).$$
$$\tag{1.25}$$

We can write the last action in the matrix form

$$g \cdot \begin{pmatrix} v \\ w \end{pmatrix} = \begin{pmatrix} 0 & \rho(g a_0) \\ \rho'(g a_0^{-1}) & 0 \end{pmatrix} \mathbb{K} \begin{pmatrix} v \\ w \end{pmatrix}. \tag{1.26}$$

The restriction and induction are adjoint functors and this is the so called Frobenius reciprocity lemma:

Lemma 1.2 (Frobenius reciprocity) *Let (G, ϕ) be a finite magnetic point group, V a representation of the core group G_0 and W a representation of (G, ϕ). Then there exist natural bijection*

$$Hom_{\mathbf{Rep}(G,\phi)} \left(Ind_{G_0}^{(G,\phi)} V, W \right) \xrightarrow{\cong} Hom_{Rep(G_0)} (V, \mathfrak{res}\, W). \tag{1.27}$$

Proof Define the functions

$$F : Hom_{\mathbf{Rep}(G,\phi)} \left(\mathbb{C}[G, \phi] \otimes_{\mathbb{C}[G_0]} V, W \right) \longrightarrow Hom_{Rep(G_0)}(V, \mathfrak{res}\, W)$$
$$f \longmapsto F(f)(v) = f(e \otimes v) \tag{1.28}$$

$$G : Hom_{Rep(G_0)}(V, \mathfrak{res}\, W) \longrightarrow Hom_{\mathbf{Rep}(G,\phi)} \left(\mathbb{C}[G, \phi] \otimes_{\mathbb{C}[G_0]} V, W \right)$$
$$f \longmapsto G(f)(g \otimes v) = g f(v) \tag{1.29}$$

F and G are inverses one of each-other:

1. if $f \in Hom_{\mathbf{Rep}(G,\phi)} \left(Ind_{G_0}^{(G,\phi)} V, W \right)$ and $g \otimes v \in Ind_{G_0}^{(G,\phi)}$ is an elementary tensor, then

$$(G \circ F)(f)(g \otimes v) = G(F(f))(g \otimes v) = g F(f)(v) = g f(e \otimes v) = f(g \otimes w). \tag{1.30}$$

2. If $h \in Hom_{Rep(G_0)}(V, \mathfrak{res}\, W)$ and $v \in V$, then

$$(F \circ G)(h)(v) = F(G(h))(v) = G(h)(e \otimes v) = e h(v) = h(v). \tag{1.31}$$

$\square$

Remark 1.2 Given any subgroup H of a magnetic point group (G, ϕ) and a representation V of H we can define an induced representation $\mathrm{Ind}_H^{(G,\phi)} V$ of (G, ϕ) and also prove the Frobenius reciprocity theorem. We need to be careful because there are two types of subgroups of (G, ϕ), hence V could be an ordinary representation or a magnetic one, and of course we have two types of restrictions and two types of inductions.

1.4 Classification of Irreducible Representations

The irreducible representations of finite magnetic groups split into three different types, one for each field $\mathbb{R}$, $\mathbb{C}$ and $\mathbb{H}$. Let us see how this is carried out.

Lemma 1.3 *Let (G, ϕ) be a finite magnetic group, $\rho : G_0 \longrightarrow GL(V)$ an irreducible representation of G_0, $a_0 \in G \backslash G_0$ and $\rho' : G_0 \longrightarrow GL(V)$ the conjugate representation by a_0. Then either:*

1. *the representations ρ and its conjugate ρ' are non-isomorphic $\rho' \not\cong \rho$, or*
2. *the representation ρ and its conjugate ρ' are isomorphic $\rho \cong \rho'$ and there exist an isomorphism $T \in \mathrm{Hom}_{\mathrm{Rep}(G_0)}(\rho', \rho)$ such that*

$$\rho(a_0^2) = \pm T\overline{T}. \tag{1.32}$$

We say the representation ρ is of **complex type** *if it satisfies the first case,* **real type** *if satisfies the second case with $\rho(a_0^2) = +T\overline{T}$ and* **quaternionic type** *the case $\rho(a_0^2) = -T\overline{T}$.*

Proof If the representations ρ and ρ' are non-isomorphic, then we are done. If ρ and ρ' are isomorphic representations, then there exists a isomorphism $S \in \mathrm{Hom}_{\mathrm{Rep}(G_0)}(a_0 V, V)$ such that

$$\rho'(g) = S^{-1}\rho(g)S \tag{1.33}$$

for all $g \in G_0$. Let us do the following computation with respect to some basis

$$\rho(a_0^{-2})\rho(g)\rho(a_0^2) = \rho(a_0^{-2} g a_0^2) \quad \text{homomorphism property} \tag{1.34}$$

$$= \rho(a_0^{-1}(a_0^{-1} g a_0)a_0) \tag{1.35}$$

$$= \overline{\rho'(a_0^{-1} g a_0)} \quad \text{definition of } \rho' \tag{1.36}$$

$$= \overline{S}^{-1} \overline{\rho(a_0^{-1} g a_0)}\, \overline{S} \quad \text{complex conjugate Eqn.(1.33)} \tag{1.37}$$

$$= \overline{S}^{-1}\rho'(g)\overline{S} \quad \text{definition of } \rho' \tag{1.38}$$

$$= \overline{S}^{-1} S^{-1}\rho(g)S\overline{S} \quad \text{Eqn.(1.33).} \tag{1.39}$$

Write the last equality as:

$$S\overline{S}\rho(a_0^{-2})\rho(g)\rho(a_0^2)\overline{S}^{-1}S^{-1} = \rho(g) \tag{1.40}$$

$$\left(S\overline{S}\rho(a_0^{-2})\right)\rho(g)\left(S\overline{S}\rho(a_0^{-2})\right)^{-1} = \rho(g) \tag{1.41}$$

so the map $S\overline{S}\rho(a_0^{-2}) \in \mathrm{End}_{\mathrm{Rep}(G_0)}(V)$ is an isomorphism of the representation ρ, and thus by the classical Schur's lemma (see for example [8, Lemma 1.7]), we have

$$S\overline{S}\rho(a_0^{-2}) = c\,\mathrm{I} \tag{1.42}$$

with $c \in \mathbb{C}^*$. Now observe that replacing $g = a_0^{-2}$ in $\rho'(g) = S^{-1}\rho(g)S$ we get

$$\rho'(a_0^{-2}) = S^{-1}\rho(a_0^{-2})S \tag{1.43}$$

and using the definition of ρ' it becomes

$$\overline{\rho(a_0^{-2})} = S^{-1}\rho(a_0^{-2})S. \tag{1.44}$$

Using these last equalities we have

$$\begin{aligned}
c\,\mathrm{I} &= S\overline{S}\rho(a_0^{-2}) & \text{Eqn.(1.42)} & \tag{1.45} \\
&= S\overline{S}\,\overline{S\rho(a_0^{-2})}S^{-1} & \text{Eqn.(1.44)} & \tag{1.46} \\
&= S\overline{c}S^{-1} & \text{Eqn.(1.42)} & \tag{1.47} \\
&= \overline{c}\,\mathrm{I} & & \tag{1.48}
\end{aligned}$$

and thus $c \in \mathbb{R}^*$. Hence there are two possibilities, either $c > 0$ or $c < 0$. Now we can define a new morphism by rescaling S by a real number

$$T := \begin{cases} \frac{1}{\sqrt{c}}S & \text{if } c > 0 \\ \frac{1}{\sqrt{-c}}S & \text{if } c < 0. \end{cases} \tag{1.49}$$

Clearly we can rewrite Eq. (1.42) as $T\overline{T} = \pm\rho(a_0^2)$. $\qquad\square$

Now we state Schur's lemma in this context.

Theorem 1.1 (Schur's lemma)
Let V, W be irreducible representations of the finite magnetic group (G, ϕ). If V and W are non-isomorphic, then $\mathrm{Hom}_{\mathbf{Rep}(G,\phi)}(V, W) = 0$. If $V \cong W$, then there are three possibilities for the restriction of V to a representation of the core group G_0:

- *$\mathrm{End}_{\mathbf{Rep}(G,\phi)}(V) \cong \mathbb{R}$, and $\mathrm{res}(V) = V$ is an irreducible representation of G_0.*
- *$\mathrm{End}_{\mathbf{Rep}(G,\phi)}(V) \cong \mathbb{C}$, and $\mathrm{res}(V) \cong V_1 \oplus V_2$ where V_1, V_2 are non-isomorphic irreducible representations of G_0 (although they are complex conjugate of each other).*

- $End_{\mathbf{Rep}(G,\phi)}(V) \cong \mathbb{H}$, and $\mathfrak{res}(V) \cong V_1 \oplus V_1$ where V_1 is an irreducible representation of G_0.

Proof Suppose V and W are non isomorphic and let $T : V \longrightarrow W$ be a morphism of these representations. Then $\ker(T) = V$ since the kernel and the image of T are subrepresentations of V and W, respectively. Therefore $T = 0$.

If $\mathfrak{res}(V)$ is an irreducible representation of G_0, then

$$\mathrm{End}_{\mathbf{Rep}(G,\phi)}(V) \subset \mathrm{End}_{\mathbf{Rep}(G_0)}(V) \cong \mathbb{C}. \tag{1.50}$$

The previous isomorphism is just the classical Schur's lemma. An endomorphism $A \in \mathrm{End}_{\mathbf{Rep}(G,\phi)}(V)$ must be of the form $A = z \cdot \mathrm{id}_V$ with $z \in \mathbb{C}$ and satisfies

$$A a_0(v) = a_0 A(v) \tag{1.51}$$
$$z \cdot \mathrm{id}_V a_0(v) = a_0 \cdot z \cdot \mathrm{id}_V(v) \tag{1.52}$$
$$z \cdot a_0(v) = \bar{z} a_0(v) \tag{1.53}$$

so $z \in \mathbb{R}$ and thus $\mathrm{End}_{\mathbf{Rep}(G,\phi)}(V) \cong \mathbb{R}$.

On the other hand, if $\mathfrak{res}(V)$ is reducible, pick any irreducible representation V_1 of G_0 in the decomposition of $\mathfrak{res}(V)$, and consider the induced representation $\mathrm{Ind}_{G_0}^{(G,\phi)} V_1 = V_1 \oplus a_0 V_1$. By the Frobenius reciprocity Theorem 1.2 we have a natural non zero morphism, induced by the inclusion $\iota : V_1 \longrightarrow \mathfrak{res}(V)$

$$\mathrm{Ind}_{G_0}^{(G,\phi)} V_1 = V_1 \oplus a_0 V_1 \longrightarrow V$$
$$(v, a_0 w) \longmapsto v + a_0 w \tag{1.54}$$

The image of this morphism is a subrepresentation of the irreducible representation V, so $V_1 \oplus a_0 V_1$ is isomorphic to V. Here we have two possibilities.

Case 1. $V_1 \cong a_0 V_1$. By Lemma 1.3, there exists an isomorphism $T : a_0 V_1 \longrightarrow V_1$ such that $T\bar{T} = -a_0^2$. If we define the morphism of representations

$$\begin{pmatrix} \mathrm{id}_{V_1} & 0 \\ 0 & T \end{pmatrix} : V_1 \oplus a_0 V_1 \longrightarrow V_1 \oplus V_1, \tag{1.55}$$

then we can transform the induced representation of (G, ϕ) on $V_1 \oplus a_0 V_1$ into a representation on $V_1 \oplus V_1$ given by the formulas

- if $g \in G_0$

$$\begin{pmatrix} \rho(g) & 0 \\ 0 & \rho(g) \end{pmatrix} = \begin{pmatrix} \mathrm{id}_{V_1} & 0 \\ 0 & T \end{pmatrix} \begin{pmatrix} \rho(g) & 0 \\ 0 & \rho'(g) \end{pmatrix} \begin{pmatrix} \mathrm{id}_{V_1} & 0 \\ 0 & T^{-1} \end{pmatrix} \tag{1.56}$$

- if $g \in G \backslash G_0$, then $g = a_0 g'$ for some $g' \in G_0$ so we only need to specify the action of a_0

$$\begin{pmatrix} 0 & -T \\ T & 0 \end{pmatrix} \mathbb{K} - \begin{pmatrix} \mathrm{id}_{V_1} & 0 \\ 0 & 1 \end{pmatrix} \begin{pmatrix} 0 & \rho(a_0^2) \\ \mathrm{id}_{V_1} & 0 \end{pmatrix} \mathbb{K} \begin{pmatrix} \mathrm{id}_{V_1} & 0 \\ 0 & T^{-1} \end{pmatrix}, \tag{1.57}$$

Now, we have the following inclusion

$$\mathrm{End}_{\mathbf{Rep}(G,\phi)}(V) \subset \mathrm{End}_{\mathbf{Rep}(G_0)}(V_1 \oplus V_1) \cong \mathbb{C}^4 \tag{1.58}$$

An endomorphism $A \in \mathrm{End}_{\mathbf{Rep}(G,\phi)}(V)$ must be of the form $A = \left(\lambda \mathrm{id}_{V_1}\ \alpha \mathrm{id}_{V_1}\ \mu \gamma \mathrm{id}_{V_1}\ \theta \mathrm{id}_{V_1}\right)$. The morphism A satisfies

$$A a_0 \begin{pmatrix} v_1 \\ v_2 \end{pmatrix} = a_0 A \begin{pmatrix} v_1 \\ v_2 \end{pmatrix} \tag{1.59}$$

$$A \begin{pmatrix} -T\mathbb{K}(v_2) \\ T\mathbb{K}(v_1) \end{pmatrix} = a_0 \begin{pmatrix} \lambda v_1 + \alpha v_2 \\ \gamma v_1 + \theta v_2 \end{pmatrix} \tag{1.60}$$

$$\begin{pmatrix} -\lambda T\mathbb{K}(v_2) + \alpha T\mathbb{K}(v_1) \\ -\gamma T\mathbb{K}(v_2) + \theta T\mathbb{K}(v_1) \end{pmatrix} = \begin{pmatrix} -\overline{\gamma} T\mathbb{K}(v_1) - \overline{\theta} T\mathbb{K}(v_2) \\ \overline{\lambda} T\mathbb{K}(v_1) + \overline{\alpha} T\mathbb{K}(v_2) \end{pmatrix} \tag{1.61}$$

So we have $\lambda = \overline{\theta}$ and $\alpha = -\overline{\gamma}$ and hence

$$\mathrm{End}_{\mathbf{Rep}(G,\phi)}(V) = \left\{ \begin{pmatrix} \lambda & \alpha \\ -\overline{\alpha} & \overline{\lambda} \end{pmatrix} \mid \lambda, \alpha \in \mathbb{C} \right\} \cong \mathbb{H} \tag{1.62}$$

Case 2. $V_1 \not\cong a_0 V_1$, so

$$\mathrm{End}_{\mathbf{Rep}(G,\phi)}(V) \subset \mathrm{End}_{\mathbf{Rep}(G_0)}(V_1 \oplus a_0 V_1) \cong \mathbb{C}^2 \tag{1.63}$$

An endomorphism $A \in \mathrm{End}_{\mathbf{Rep}(G,\phi)}(V)$ must be of the form $A = \begin{pmatrix} \lambda\mathrm{id} & 0 \\ 0 & \alpha\mathrm{id} \end{pmatrix}$ and it satisfies

$$A a_0 \begin{pmatrix} v \\ a_0 w \end{pmatrix} = a_0 A \begin{pmatrix} v \\ a_0 w \end{pmatrix} \tag{1.64}$$

$$A \begin{pmatrix} a_0^2 w \\ a_0 v \end{pmatrix} = a_0 \begin{pmatrix} \lambda v \\ \alpha a_0 w \end{pmatrix} \tag{1.65}$$

$$\begin{pmatrix} \lambda a_0^2 w \\ \alpha a_0 v \end{pmatrix} = \begin{pmatrix} \overline{\alpha} a_0^2 w \\ \overline{\lambda} a_0 v \end{pmatrix} \tag{1.66}$$

so $\lambda = \overline{\alpha}$ and hence

$$\mathrm{End}_{\mathbf{Rep}(G,\phi)}(V) = \left\{ \begin{pmatrix} \lambda & 0 \\ 0 & \overline{\lambda} \end{pmatrix} \mid \lambda \in \mathbb{C} \right\} \cong \mathbb{C}. \tag{1.67}$$

$$\sqcup$$

On the other hand, if we start with an irreducible representation V of G_0, then Wigner [7] showed us a procedure to produce irreducible representations of the magnetic group of (G, ϕ).

Theorem 1.2 (Wigner) *If (G, ϕ) is a finite magnetic group and $\rho : G_0 \longrightarrow GL(\mathbb{C}^n)$ an irreducible representation of the core group G_0, then we can construct an irreducible representation of (G, ϕ), in the following way. Let $\rho' : G_0 \longrightarrow GL(a_0\mathbb{C}^n)$ be the conjugate representation of ρ,*

- **Real type**. *If ρ and ρ' are isomorphic representations $\rho'(g) = T^{-1}\rho(g)T$ with $T\overline{T} = \rho(a_0^2)$, then for $v \in \mathbb{C}^n$ the formula*

$$g \cdot v := \begin{cases} \rho(g)v & g \in G_0 \\ \rho(ga_0^{-1})T\mathbb{K}(v) & g \in G \backslash G_0 \end{cases} \tag{1.68}$$

defines an irreducible representation of (G, ϕ).
- **Complex type**. *If ρ and ρ' are non-isomorphic representations, then for $v, w \in \mathbb{C}^n$ the formula*

$$g \cdot (v, w) := \begin{cases} \begin{pmatrix} \rho(g) & 0 \\ 0 & \rho'(g) \end{pmatrix} \begin{pmatrix} v \\ w \end{pmatrix} & g \in G_0 \\ \begin{pmatrix} 0 & \rho(ga_0) \\ \rho'(ga_0^{-1}) & 0 \end{pmatrix} \mathbb{K} \begin{pmatrix} v \\ w \end{pmatrix} & g \in G \backslash G_0 \end{cases} \tag{1.69}$$

*defines an **irreducible representation of** (G, ϕ).*
- **Quaternionic type**. *If ρ and ρ' are isomorphic representations $\rho'(g) = T^{-1}\rho(g)T$ with $T\overline{T} = -\rho(a_0^2)$, then for $v, w \in \mathbb{C}^n$ the formula*

$$g \cdot (v, w) := \begin{cases} \begin{pmatrix} \rho(g) & 0 \\ 0 & \rho(g) \end{pmatrix} \begin{pmatrix} v \\ w \end{pmatrix} & g \in G_0 \\ \begin{pmatrix} 0 & \rho(ga_0^{-1})T \\ -\rho(ga_0^{-1})T & 0 \end{pmatrix} \mathbb{K} \begin{pmatrix} v \\ w \end{pmatrix} & g \in G \backslash G_0 \end{cases} \tag{1.70}$$

*defines an **irreducible representation of** (G, ϕ).*

Moreover, every irreducible representation of (G, ϕ) is constructed in this way.

The first proof of this theorem was given by Wigner in Sects. 4 and 5 of Chap. 20 in [7].

Proof Consider the induced representation $\mathrm{Ind}_{G_0}^{(G,\phi)} \rho$. First, suppose it is irreducible. We have $\mathrm{resInd}_{G_0}^{(G,\phi)} \rho = \rho \oplus \rho'$, so by the Schur's Lemma 1.1 the representation ρ can be of complex or quaternionic type. If ρ is of complex type, then ρ and ρ' are non-isomorphic representations and the action described at the end of Definition 1.4, see Eqs. (1.24) and (1.26), give us the formula of the theorem. If ρ is of quaternionic type, then ρ and ρ' are isomorphic and the representation $\mathrm{Ind}_{G_0}^{(G,\phi)}\rho$ take the form of described in this theorem by performing the change of basis given in the proof of Theorem 1.1.

If $\mathrm{Ind}_{G_0}^{(G,\phi)}\rho$ is reducible, then ρ is of real type and $\mathrm{Ind}_{G_0}^{(G,\phi)}\rho$ contains to copies of the representation described in this theorem.

Finally, if ρ is an irreducible representation of (G,ϕ), then by Theorem 1.1 the representation $\mathfrak{res}\,\rho$ is an irreducible representation or a sum of two of irreducible representations of the core group G_0. In any case, we see that we can take these pieces of $\mathfrak{res}\,\rho$ and reconstruct ρ as this theorem states. $\square$

Remark 1.3 Let ρ be an irreducible representation of (G,ϕ). By Theorem 1.2 ρ comes from an irreducible representation of the core group G_0. We also say that ρ is of **real**, **complex**, or **quaternion type** if it comes from an irreducible representation of the core of the respectively same type.

Corollary 1.1 *For any representation V of a finite magnetic group (G,ϕ), there is a decomposition*

$$V = \left(A_1^{a_1} \oplus \ldots \oplus A_m^{a_m}\right) \oplus \left(B_1^{b_1} \oplus \ldots \oplus B_n^{b_n}\right) \oplus \left(C_1^{c_1} \oplus \ldots \oplus C_l^{c_l}\right) \tag{1.71}$$

where A_i, B_j, C_k are non-isomorphic irreducible representations of real, complex and quaternionic type, respectively.

If we denote by $\mathbf{R}(G,\phi;\mathbb{F})$ the free abelian group generated by the irreducible representations of real, complex, quaternionic type for $\mathbb{F} = \mathbb{R}, \mathbb{C}, \mathbb{H}$, respectively, then the representation group of (G,ϕ) splits as

$$\mathbf{R}(G,\phi) \cong \mathbf{R}(G,\phi;\mathbb{R}) \oplus \mathbf{R}(G,\phi;\mathbb{C}) \oplus \mathbf{R}(G,\phi;\mathbb{H}) \tag{1.72}$$

and by the Theorem 1.1

$$\mathrm{End}_{\mathbf{Rep}(G,\phi)}(V) \cong \bigoplus_i \mathrm{End}_{\mathbf{Rep}(G,\phi)}(A_i)^{a_i} \oplus \bigoplus_j \mathrm{End}_{\mathbf{Rep}(G,\phi)}(B_j)^{b_j} \oplus \bigoplus_k \mathrm{End}_{\mathbf{Rep}(G,\phi)}(C_k)^{c_k}$$

$$\cong \bigoplus_i \mathbb{R}^{a_i} \oplus \bigoplus_j \mathbb{C}^{b_j} \oplus \bigoplus_k \mathbb{H}^{c_k}.$$

Given an irreducible representation of the core $\rho : G_0 \longrightarrow \mathrm{GL}(n)$, then there is a procedure to find out which type of irreducible representation of (G,ϕ) will produce. This is called the Schur-Frobenius indicator. Denote by χ_ρ the character of ρ and fix an element $a_0 \in G\backslash G_0$.

Definition 1.5 The **Schur-Frobenius indicator** of ρ with respect to ϕ is

$$\mathrm{SchurF}_\phi(\rho) = \sum_{g \in a_0 G_0} \chi_\rho(g^2) \tag{1.74}$$

Notice this number is well defined since for every $g \in a_0 G_0$, $g^2 \in G_0$.

Theorem 1.3 *Let (G, ϕ) be a finite magnetic group and $\rho : G_0 \longrightarrow GL(n)$ an irreducible representation of G_0.*

- *If $SchurF_\phi(\rho) = |G_0|$, then the irreducible representation of (G, ϕ) associated to ρ is of real type.*
- *If $SchurF_\phi(\rho) = 0$, then the irreducible representation of (G, ϕ) associated to ρ is of complex type.*
- *If $SchurF_\phi(\rho) = -|G_0|$, then the irreducible representation of (G, ϕ) associated to ρ is of quaternionic type.*

Proof We refer the reader to [9, sect. 7] for the proof of the result, which is essentially based on the orthogonally relations between characters. $\qquad\qquad\square$

Lemma 1.4 *For $\mathbb{F} \cong \mathbb{R}$, $\mathbb{C}$ or $\mathbb{H}$, let V, $W \in Irrep(G, \phi; \mathbb{F})$ and E a finite $\mathbb{F}$-vector space. Then we have the following isomorphism of $\mathbb{F}$-vector spaces*

$$Hom_{\mathbf{Rep}(G,\phi)}(V, W \otimes_{\mathbb{F}} E) \cong Hom_{\mathbf{Rep}(G,\phi)}(V, W) \otimes_{\mathbb{F}} E \qquad (1.75)$$

Proof

$$Hom_{\mathbf{Rep}(G,\phi)}(V, W \otimes_{\mathbb{F}} E) \cong Hom_{\mathbf{Rep}(G,\phi)}(V, W \otimes_{\mathbb{F}} \mathbb{F}^n) \qquad (1.76)$$
$$\cong Hom_{\mathbf{Rep}(G,\phi)}(V, W^n) \qquad (1.77)$$
$$\cong Hom_{\mathbf{Rep}(G,\phi)}(V, W)^n \qquad (1.78)$$
$$\cong Hom_{\mathbf{Rep}(G,\phi)}(V, W) \otimes_{\mathbb{F}} \mathbb{F}^n \qquad (1.79)$$
$$\cong Hom_{\mathbf{Rep}(G,\phi)}(V, W) \otimes_{\mathbb{F}} E. \qquad (1.80)$$

$$\square$$

References

1. Lifshitz, R.: Magnetic point groups and space groups. In: Bassani, F., Liedl, G.L., Wyder, P. (eds.), Encyclopedia of Condensed Matter Physics, pp. 219–226. Elsevier, Oxford (2005)
2. Zhang, F., Kane, C.L., Mele, E.J.: Time-reversal-invariant topological superconductivity and majorana kramers pairs. Phys. Rev. Lett. **111** 056402 (2013)
3. Fu, L., Kane, C.L., Mele, E.J.: Topological insulators in three dimensions. Phys. Rev. Lett. **98**, 106803 (2007)
4. Bradley, C.J., Cracknell, A.P.: The mathematical theory of symmetry in solids. Oxford Classic Texts in the Physical Sciences. The Clarendon Press, Oxford University Press, New York, paperback edition (2010). Representation theory for point groups and space groups
5. Atiyah, M.F., Segal, G.B.: Equivariant K-theory and completion. J. Diff. Geom. **3**, 1–18 (1969)
6. Atiyah, M.F.: K-theory and reality. Quart. J. Math. Oxford Ser. **2**(17), 367–386 (1966)
7. Wigner, E.P.: Group theory and its application to the quantum mechanics of atomic spectra, volume, Vol. 5 of Pure and Applied Physics. Academic Press, New York-London: Expanded and, improved edn. Translated from the German by J. J. Griffin (1959)

8. Fulton, W., Harris, J.: Representation theory, volume 129 of Graduate Texts in Mathematics. Springer, New York, (1991). A first course, Readings in Mathematics
9. Newmarch, J.D., Golding, R.M.: The character table for the corepresentations of magnetic groups. J. Math. Phys. **23**(5), 695–704 (1982)
10. Bárcenas, J.E., Joachim, M., Uribe, B.: Universal twist in equivariant K-theory for proper and discrete actions. Proc. Lond. Math. Soc. (3) **108**(5), 1313–1350 (2014)

Chapter 2
Magnetic Equivariant K-theory

Abstract In this chapter we define magnetic equivariant vector bundles and, via the Grothendieck construction, the magnetic equivariant K-theory groups. We show that the magnetic equivariant K-theory defines an equivariant cohomology theory, calculate its coefficients, we show the Bott periodicity, the Thom isomorphism, the degree shift isomorphism and the Atiyah-Hirzebruch spectral sequence.

Keywords Magnetic vector bundle · Magnetic K-theory · Real K-theory · Symplectic K-theory · Equivariant K-theory · Degree shift · Twistings

In this chapter we will develop the K-theory for spaces with actions of magnetic groups. This is a generalization of Atiyah's Real K-theory [1], Segal's equivariant K-theory [2] and some twisted K-theories [3]. This is not the first time such a notion is introduced. For instance, M. Karoubi was the first to define it in [4]. More recently, it has been articulated with the study of topological phases of matter [5, 6]. In these works D. Freed, G. Moore and K. Gomi defined the so-called Freed-Moore K-theory, that essentially contains the definition we will state next. This particular K-theory has been further developed to provide tools for its explicit calculation [7, 8].

We carry out several computations, mainly for tori equipped with actions of the magnetic point groups studied in the previous chapter; in particular we compute the coefficients of this K-theory and, as one can expect, the representations of the magnetic point groups will appear. The way we compute these groups is via Clifford modules. These results can be potentially used to detect protected topological phases of matter, and in the near future, they could be used to produce periodic tables for topological insulators and superconductors, generalizing the work of Kitaev [9].

Moreover, we will discuss the role of extensions of magnetic point groups by subgroups of U(1), which naturally lead to what are known as twisted representations. These twisted representations are particularly relevant in cases where the time-reversal operator no longer squares to the identity but instead to minus the identity. This occurs, for instance, in systems with Spin-Orbit Coupling, where time-reversal symmetry lifts to a group of order 4, [10]. The spatial symmetries, which often reside in SO(3), must then be lifted to its double cover, SU(2), to accommodate spin-1/2 particles in quantum systems. This modification affects not only the

B. Uribe Jongbloed et al., *Magnetic Equivariant K-Theory*,
SpringerBriefs in Mathematics, https://doi.org/10.1007/978-3-032-05914-7_2

algebraic structure of the symmetry group but also how representations of the group relate to physical observables, such as energy levels and degeneracies.

2.1 Magnetic Equivariant Vector Bundles

For a magnetic group (G, ϕ) the category of (G, ϕ)-spaces will simply be the category of G-spaces. Some explicit choices of magnetic groups have been important in the literature, let us see some of them.

A $(\mathbb{Z}/2, \mathrm{id})$-space X is the same as a Real space in the sense of Atiyah [1], i.e. a space with involution given by the action of the non trivial element of $\mathbb{Z}/2$. Recall, an *involution* on X is a homeomorphism $\tau : X \longrightarrow X$ such that $\tau^2 = \mathrm{id}$.

A $(G_0 \rtimes \mathbb{Z}/2, \pi_2)$-space X is the same as a Real G_0-equivariant space in the sense of Atiyah and Segal [2], that is, a G_0-space X with an involution τ such that $\tau(g \cdot x) = \tau(g) \cdot \tau(x)$ for $g \in G_0$ and $x \in X$, and where $\tau : G_0 \longrightarrow G_0$ denotes the action of the non-trivial element of $\mathbb{Z}/2$ on G_0.

The following spaces are (G, ϕ)-spaces via the homomorphism $G \xrightarrow{\phi} \mathbb{Z}/2$, where the non trivial element of $\mathbb{Z}/2$ acts by complex conjugation:

- $\mathbb{R}^{p,q} = \mathbb{R}^q \oplus i\mathbb{R}^p$ with the involution by complex conjugation $x + iy \longmapsto x - iy$.
- $D^{p,q}$ unit ball in $\mathbb{R}^{p,q}$ (with the euclidean norm on $\mathbb{R}^{p+q}$).
- $S^{p,q}$ unit sphere in $\mathbb{R}^{p,q}$.

The interchange on the super indices position is due to Atiyah's convention in [1].

Let V be a complex vector space, then

$$P(V) = (V \setminus \{0\})/\mathbb{C}^* \tag{2.1}$$

is the projective space over V. If V is a representation of (G, ϕ), then $P(V)$ is a (G, ϕ)-space with the action $g \cdot [v] := [gv]$. This action is well-defined because $g(\lambda v) = \mathbb{K}^{\phi(g)}(\lambda)gv$ and therefore $g[\lambda v] = g[v]$.

The vector bundles relevant to our analysis in this work are described below.

Definition 2.1 A **magnetic (G, ϕ)-equivariant vector bundle** is a *complex vector bundle* $E \xrightarrow{p} X$ over a compact G-space X such that

- the total space E is a G-space and the projection p is G-equivariant.
- For $g \in G$ and $x \in X$ the fiberwise map $g : E_x \longrightarrow E_{g \cdot x}$ is complex linear if $\phi(g) = 0$ and complex antilinear if $\phi(g) = 1$, for all $x \in X$.

Morphisms of magnetic (G, ϕ)-equivariant vector bundles are morphisms of complex vector bundles $f : E \longrightarrow F$ commuting with the action of G.

Denote by $\mathbf{Vect}_{(G,\phi)}(X)$ the **category of magnetic (G,ϕ)-equivariant vector bundles**. Denote the vector space of morphisms between E and F by $\mathrm{Hom}_{(G,\phi)}(E, F)$.

A **section** of a (G,ϕ)-vector bundle is a G-equivariant section $s : X \longrightarrow E$. Denote the vector space of such sections as $\Gamma_G(E)$.

Construction of magnetic (G,ϕ)-equivariant vector bundles.
Let E, F be two (G,ϕ)-vector bundles over X.

- The **hom bundle** $\mathrm{Hom}(E, F)$ with fiber over x the complex vector space of linear transformations $\mathrm{Hom}(E_x, F_x)$ between E_x and F_x, and action of (G, ϕ) given by

$$(g \cdot f)(v) = gf(g^{-1}v)$$

  for $g \in G$, $f \in \mathrm{Hom}(E, F)$ and $v \in E_{gx}$.
- The **tensor bundle** $E \otimes F$ with fiber $E_x \otimes F_x$ at $x \in X$ and action of (G, ϕ) given by

$$g \cdot (v \otimes w) = gv \otimes gw$$

  for $g \in G$, $v \in E_x$ and $w \in F_x$.
- The **pull-back** f^*E for $f : Y \longrightarrow X$ a G-equivariant map of G-spaces with the induced action of (G, ϕ).
- If $X = X_1 \cup X_2$ is the union of two compact (G, ϕ)-subspaces and $A = X_1 \cap X_2$ is a compact (G, ϕ)-subspace, the **clutching** of two (G, ϕ)-vector bundles $E_1 \longrightarrow X_1$ and $E_2 \longrightarrow X_2$ via a (G, ϕ)-isomorphism $\alpha : E_1|_A \longrightarrow E_2|_A$ is given by

$$E_1 \cup_\alpha E_2 := E_1 \sqcup E_2 \big/ \left(e \sim \alpha(e) \text{ for all } e \in (E_1)_x \text{ with } x \in A \right)$$

With all the previous notation, we can state the following classical fact:

Lemma 2.1 *Let E, F be two (G, ϕ)-vector bundle over X. Then we have a natural isomorphism*

$$\mathrm{Hom}_{(G,\phi)}(E, F) \xrightarrow{\ \cong\ } \Gamma_G(\mathrm{Hom}(E, F))$$
$$f \longmapsto s_f$$

between morphisms of (G, ϕ)-vector bundles and G-equivariant sections of the bundle $\mathrm{Hom}(E, F)$ over X. Here $s_f(x)(v) = f(v)$ for all $v \in E_x$.

Example 2.1 Let n be a non-negative integer and X a (G, ϕ)-space. The **trivial (G, ϕ)-vector bundle over X of dimension** n is

$$\mathbf{n} := X \times \mathbb{C}^n$$
$$\downarrow{\scriptstyle \pi_1}$$
$$X$$

with the trivial action

$$g \cdot (x, z) = \left(gx, \mathbb{K}^{\phi(g)}(z)\right)$$

where $g \in G$, $x \in X$ and $z \in \mathbb{C}^n$. More generally, let V be a representation of (G, ϕ), then $\mathbf{V} := X \times V \xrightarrow{\pi_1} X$ is the **trivial bundle** over X with fiber V.

Definition 2.2 (*Stable equivalence*) Two (G, ϕ)-vector bundles E, F over a G-space X are **stably equivalent** $E \sim_s E'$ if there exist trivial (G, ϕ)-vector bundles $\mathbf{V}$, $\mathbf{V}'$ on X such that $E \oplus \mathbf{V} \cong E' \oplus \mathbf{V}'$.

Remark 2.1 Notice the set of stable equivalence classes of (G, ϕ)-vector bundles is an abelian semigroup under $\oplus$, whose identity is given by the class of the trivial bundle. However, it is not clear that this is an abelian group since we do not yet know if for every (G, ϕ)-vector bundle E there exists another F such that $E \oplus F \cong \mathbf{V}$. In fact, we will show this at the end of this section in Proposition 2.2.

2.2 Definition of Magnetic Equivariant K-Theory

Definition 2.3 The **magnetic equivariant** K-**theory** $\mathbf{K}_G(X)$ of the G-space X is the Grothendieck group of the category $\mathbf{Vect}_{(G,\phi)}(X)$ of magnetic (G, ϕ)-equivariant vector bundles over X. Its elements are formal differences $E_0 - E_1$ of magnetic (G, ϕ)-equivariant vector bundles, modulo the equivalence relation $E_0 - E_1 \sim E_0' - E_1'$ if and only if $E_0 \oplus E_1' \oplus F \cong E_0' \oplus E_1 \oplus F$ for some magnetic (G, ϕ)-equivariant vector bundle F on X.

Remark 2.2 From now on, we will remove the homomorphism ϕ from the notation. Boldface functors receive magnetic groups and ϕ will be assumed to be part of the structure of the magnetic groups. Hence, we will denote by

$$\mathbf{Rep}(G), \ \mathbf{R}(G), \ \mathbf{R}(G, \mathbb{F}), \ \mathbf{Irrep}(G, \mathbb{F}), \ \mathbf{Vect}_G(X) \text{ and } \mathbf{K}_G(X) \qquad (2.2)$$

the category of magnetic (G, ϕ)-representations, its Grothendieck group, the subgroup generated by irreducible representations of type $\mathbb{F}$ (for $\mathbb{F} \in \{\mathbb{R}, \mathbb{C}, \mathbb{H}\}$), the set of isomorphism classes of irreducible representations of type $\mathbb{F}$, the category of magnetic (G, ϕ)-equivariant vector bundles and its K-theory respectively.

Example 2.2 If the magnetic group is $(\mathbb{Z}/2, \mathrm{id})$, then a $(\mathbb{Z}/2, \mathrm{id})$-vector bundle is a Real vector bundle in the sense of Atiyah [1]. We have then that magnetic $(\mathbb{Z}/2, \mathrm{id})$-equivariant K-theory is equivalent to Atiyah's Real K-theory,

$$\mathbf{K}_{\mathbb{Z}/2}(X) = K\mathbb{R}(X), \qquad (2.3)$$

and therefore, if X is a trivial $\mathbb{Z}/2$ space, then $\mathbf{K}_{\mathbb{Z}/2}(X) = KO(X)$.

Example 2.3 Consider magnetic group

$$1 \longrightarrow G_0 = \mathbb{Z}/2 \longrightarrow G = \mathbb{Z}/4 \xrightarrow{\text{mod}\,2} \mathbb{Z}/2 \longrightarrow 1 \tag{2.4}$$

together with a space X with an involution. Take the induced
   action of $\mathbb{Z}/4$ on X given by the homomorphism $\phi := \text{mod}\,2$.
   Since the core $G_0 = \mathbb{Z}/2$ acts trivially on X, any magnetic $(\mathbb{Z}/4, \phi)$-equivariant
vector bundle $E \to X$ can be split as $E \cong E_1 \oplus E_{-1}$ where G_0 acts trivially on E_1
and with the sign representation on E_{-1}. If τ denotes the generator of $\mathbb{Z}/4$, then τ acts
complex antilinearly on E, with $\tau^2 = 1$ on E_1 and with $\tau^2 = -1$ on E_{-1}. Therefore,
E_1 is a Real vector bundle in the sense of Atiyah [1] and E_{-1} is a quaternionic
bundle [11]. Hence for X a $\mathbb{Z}/2$-space we have the canonical isomorphism

$$\mathbf{K}_{\mathbb{Z}/4}(X) \cong K\mathbb{R}(X) \oplus K\mathbb{H}(X), \tag{2.5}$$

that in the case that X is a trivial $\mathbb{Z}/2$ space becomes

$$\mathbf{K}_{\mathbb{Z}/4}(X) \cong K O(X) \oplus K Sp(X) \tag{2.6}$$

where $K Sp$ is symplectic K-theory as defined by Dupont [12].

Let us denote by $K U$ the usual complex K-theory.

Example 2.4 Suppose we take the space X to be a point $*$ and G finite. Then
any magnetic (G, ϕ)-equivariant vector bundle over a point is nothing else as a
representation of the magnetic group (G, ϕ). Then we get the canonical isomorphism
of groups

$$\mathbf{K}_G(*) \cong \mathbf{R}(G). \tag{2.7}$$

This generalizes what happens in the equivariant case as shown by Segal [13].
   Now we will carry out some computations which are based on the constructions
of some induced bundles.

Definition 2.4 Let (G, ϕ) be a magnetic group, $H \leq G$ a subgroup and V a complex
representation of the group H. The **induced bundle** is the (G, ϕ)-vector bundle over
the discrete space G/H

$$p : G \times_H V \longrightarrow G/H \tag{2.8}$$

where

- $G \times_H V = G \times V / \big((gh^{-1}, hv) \sim (g, v) \text{ for } h \in H\big)$ with the quotient topology,
- the projection p is given by $p([g, v]) = gH$,
- the vector space structure on each fiber is

$$[g, v] + [g, w] = [g, v + w], \quad z[g, v] = [g, \mathbf{K}^{\phi(g)}(z)v] \qquad (2.9)$$

for $g \in G$, $v, w \in V$, and $z \in \mathbb{C}$.

- and the action of (G, ϕ) on $G \times_H V$ is $l \cdot [g, v] = [lg, v]$.

Note that the action of (G, ϕ) on $G \times_H V$ is linear or antilinear depending on ϕ:

$$\begin{aligned}
l \cdot (z[g, v]) &= [lg, \mathbb{K}^{\phi(g)}(z)v] \\
&= \mathbb{K}^{\phi(l)}(z)[lg, v] \\
&= \mathbb{K}^{\phi(l)}(z)l[g, v].
\end{aligned}$$

In the particular case where $H = G_0$ and $a_0 \in G \backslash G_0$ is fixed, then we have an isomorphism

$$G \times_{G_0} V \longrightarrow V \sqcup a_0 V$$

$$[g, v] \longmapsto \begin{cases} a_0(a_0^{-1}gv) & \text{if } g \notin G_0 \\ gv & \text{if } g \in G_0 \end{cases}$$

where $a_0 V = \{a_0 v \mid v \in V\}$ is the conjugated representation of V introduced in Eqs. (1.21) and (1.22).

Lemma 2.2 *There is a natural isomorphism*

$$\mathbf{K}_G(S^{1,0}) \cong KU_{G_0}(*) \cong \mathrm{R}(G_0).$$

Proof Let $p : E \longrightarrow \{\pm 1\}$ be a (G, ϕ)-vector bundle with $S^{1,0} \cong \{\pm 1\}$ with the free $\mathbb{Z}/2$ and the G action induced by ϕ. Then the restriction $p^{-1}(\{1\}) \longrightarrow \{1\} = *$ is a G_0-equivariant vector bundle.

On the other hand, if $V \longrightarrow *$ is a G_0-equivariant vector bundle, then consider the induced bundle

$$p : E = (G, \phi) \times_{G_0} V \longrightarrow \{\pm 1\}$$

These constructions are inverses of each other. $\square$

We can generalize the previous lemma by taking any subgroup H of G.

Lemma 2.3 *Let (G, ϕ) be a finite magnetic group and H a subgroup of G. Then there is a canonical isomorphism*

$$\mathbf{K}_G(G/H) \cong \begin{cases} KU_H(*) & \text{if } H \leq G_0 \\ \mathbf{K}_H(*) & \text{if } H \nleq G_0 \end{cases}$$

Proof The proof uses the same constructions as in the previous lemma. Let $p : E \longrightarrow G/H$ be a (G, ϕ)-vector bundle. Then the restriction $E|_{\{H\}} \longrightarrow \{H\} = *$ is an H- or $(H, \phi|_H)$-vector bundle over the point, depending on whether $H \leq G_0$ or $H \not\leq G_0$.

Conversely, let $V \longrightarrow *$ be an H or $(H, \phi|_H)$-vector bundle and consider the induced bundle

$$p : E = G \times_H V \longrightarrow G/H.$$

These constructions are inverses of each other. $\qquad\square$

Proposition 2.1 *Let (G, ϕ) be a finite magnetic group acting trivially on X. Then we have a natural isomorphism of groups:*

$$\mathbf{K}_G(X) \xrightarrow{\cong} (\mathbf{R}(G, \mathbb{R}) \otimes KO(X)) \oplus (\mathbf{R}(G, \mathbb{C}) \otimes KU(X)) \oplus (\mathbf{R}(G, \mathbb{H}) \otimes KSp(X)).$$
$$(2.10)$$

Proof Take a magnetic (G, ϕ)-equivariant vector bundle $p : E \longrightarrow X$ and $V \in$ **Irrep**$(G, \mathbb{F})$. The bundle

$$\mathrm{Hom}_{\mathbf{Rep}(G,\phi)}(V, E) \to X \qquad (2.11)$$

is an $\mathrm{End}_{\mathbf{Rep}(G,\phi)}(V) \cong \mathbb{F}$-vector bundle, and therefore, the map

$$\mathbf{K}_G(X) \xrightarrow{\cong} \bigoplus_{\mathbb{F}\in\{\mathbb{R},\mathbb{C},\mathbb{H}\}} \mathbf{R}(G, \mathbb{F}) \otimes K\mathbb{F}(X) \qquad (2.12)$$

$$E \longmapsto \bigoplus_{\mathbb{F}\in\{\mathbb{R},\mathbb{C},\mathbb{H}\}} \bigoplus_{V\in\mathbf{Irrep}(G,\mathbb{F})} V \otimes \mathrm{Hom}_{\mathbf{Rep}(G,\phi)}(V, E)$$

provides the desired isomorphism. The inverse map takes a representation V of type $\mathbb{F}$ and an $\mathbb{F}$-vector bundle E in $K\mathbb{F}(X)$ to $V \otimes_{\mathbb{F}} E$ in $\mathbf{K}_G(X)$. $\qquad\square$

2.3 Topological Properties

Now we prove some lemmas that are required to extend $\mathbf{K}_G$ to a cohomology theory; these are extensions of the lemmas appearing in [1, 13–16]. For the rest of the section we assume:

- X is a compact G-CW complex (see Definition 2.19).
- Magnetic point groups (G, ϕ) are finite.

In [16, Theorem 14.2] it is shown that a compact G-CW complex X is also a G-absolute neighborhood retract (G-ANR), that is, X is metrizable and, whenever X is a closed G-subspace of a metrizable G-space Y, X is a G-neighborhood retract of Y. The existence of extensions presented in the next lemmas follows directly from this fact. We include small proofs for completeness.

Lemma 2.4 *Let X be a G-space, $A \subset X$ a closed invariant subspace and W an open neighborhood of A. Then there exist an open invariant set U such that $A \subset U \subset W$.*

Proof Define the open set

$$U := q^{-1}\left(X/G \setminus q(X \setminus W)\right)$$

where $q : X \longrightarrow X/G$ is the canonical projection. First, note that $q(A) \cap q(X \setminus W) = \emptyset$: Suppose $[x] \in q(A) \cap q(X \setminus W) = \emptyset$, then there exist $g, h \in G$ such that $g \cdot x \in A$ and $h \cdot x \in X \setminus W$. So $h \cdot (hg^{-1}) \cdot g \cdot x \in A$ since A is invariant. Then $h \cdot x \in A \cap X \setminus W$ which is a contradiction. So we have $q(A) \subset X/G \setminus q(X \setminus W)$ and thus $A = q^{-1}(q(A)) \subset q^{-1}(X/G \setminus q(X \setminus W)) = U$ since A is invariant. Now, $U \subset W$: suppose there exist $x \in U$ such that $x \notin W$; then $[x] \in p(U) = X/G \setminus q(X \setminus W)$ and $[x] \in q(X \setminus W)$ which is a contradiction. $\qquad\square$

This proof is based in the one given by Palais [17, Sect. 1.1.14].

Lemma 2.5 (Extension of sections) *If E is a (G, ϕ)-vector bundle on a G-space X, and A is a closed G-subspace of X, then any section of the (G, ϕ)-vector bundle $E|_A$ can be extended to a section of the (G, ϕ)-vector bundle E.*

Proof Let $s : A \longrightarrow E|_A$ be such a section and extend it arbitrarily to E, as in [14, Sect. 1.4.1] or by using the Tietze extension theorem to extend it locally and then add this local extension to a global section via a partition of the unity. Finally, take the average of s over G

$$s(x) = \frac{1}{|G|} \sum_G (g \cdot s)(x)$$

as in [13, Sect. 1.1] or [1, Sect. 1]. $\qquad\square$

Lemma 2.6 *Let E, F be two (G, ϕ)-vector bundles on X, A a closed G-subspace of X and $f : E|_A \longrightarrow F|_A$ an isomorphism. Then there is a G-neighborhood U of A in X and an isomorphism $f : E|_U \longrightarrow F|_U$ of (G, ϕ)-vector bundles extending f.*

Proof By Lemma 2.1, $f|_A$ can be understood as a section of the (G, ϕ)-vector bundle $\mathrm{Hom}_{(G,\phi)}(E|_A, F|_A) = \mathrm{Hom}_{(G,\phi)}(E, F)|_A$ over A. Using Lemma 2.5, extend this section to a section f of $\mathrm{Hom}_{(G,\phi)}(E, F)$. So we have a morphism from E to F. The subset W of X where f is an isomorphism, is an open set which contains A by the continuity of the determinant. By Lemma 2.4 there exists a G-equivariant open neighborhood U of A contained in W, and of course, f is still an isomorphism on $E|_U$. $\qquad\square$

Lemma 2.7 *Let $p : E \longrightarrow X$ be a (G, ϕ)-vector bundle, $f_0, f_1 : Y \longrightarrow X$ be two G-homotopic G-maps and Y be a compact space. Then we have an isomorphism $f_0^* E \cong f_1^* E$ of (G, ϕ)-vector bundles.*

Proof Suppose $H : Y \times I \longrightarrow X$ is such an equivariant homotopy between f_0 and f_1. For any $t \in I$, we have an isomorphism of (G, ϕ)-vector bundles

$$H^* E|_{Y \times \{t\}} \longrightarrow \pi_t^* H_t^* E|_{Y \times \{t\}} = H_t^* E \tag{2.13}$$

over $Y \times \{t\}$, where $\pi_t : Y \times I \longrightarrow Y \times \{t\}$ is given by $\pi_t(y, s) = (y, t)$ and $H_t = H|_{y \times \{t\}}$. By Lemma 2.6 there is an open neighborhood of $Y \times \{t\}$, namely $Y \times \delta(t)$, where $H^* E$ and $\pi_1^* H_t^* E$ are isomorphic. If $r \in \delta(t)$, then we have a commutative diagram:

$$
\begin{array}{ccc}
(H^* E)|_{Y \times \delta(t)} & \xrightarrow{\;\;\cong\;\;} & (\pi_t^*(H_t^* E))|_{Y \times \delta(t)} \\
\uparrow & & \uparrow \\
H_r^* E = (H^* E|_{Y \times \delta(t)})|_{Y \times \{r\}} & \dashrightarrow{\;\cong\;} & (\pi_t^*(H_t^* E)|_{Y \times \delta(t)})|_{Y \times \{r\}}
\end{array}
\tag{2.14}
$$

The vector bundle $\pi_t^* H_t^* E|_{Y \times \delta(t)}|_{Y \times \{r\}}$ is the same vector bundle as $\theta_{t,r}^* H_t^* E$ where $\theta_{t,r} : Y \times \{r\} \longrightarrow Y \times \{t\}$ is given by $\theta_{t,r}(y, r) = (y, t)$ and also we have a natural isomorphism $\theta_{t,r}^* H_t^* E \cong H_t^* E$. Hence, the isomorphism class of $H_t^* E$ is locally constant function of t. Since the interval I is connected, this implies it is constant. This is the same argument given in [14, Lemma 1.4.3]. $\qquad\square$

Corollary 2.1 *If $f : X \longrightarrow Y$ is a G-homotopy equivalence, then the pull-back*

$$f^* : \mathbf{Vect}_{(G,\phi)}(Y)/\cong \; \longrightarrow \; \mathbf{Vect}_{(G,\phi)}(X)/\cong \tag{2.15}$$

is a bijection between the isomorphism classes of (G, ϕ)-vector bundles over Y and X. In particular, if X is G-contractible, then every (G, ϕ)-vector bundle over X is isomorphic to a trivial (G, ϕ)-vector bundle.

Let $X = X_1 \cup X_2$, $A = X_1 \cap X_2$, all spaces being compact and G-invariant. Assume that E_i is a (G, ϕ)-vector bundles over X_i and that $\alpha : E_1|_A \longrightarrow E_2|_A$ is an (G, ϕ)-isomorphism. The **clutching** of E_1 and E_2 is the following (G, ϕ)-vector bundle over X

$$E_1 \cup_\alpha E_2 := E_1 \sqcup E_2/(v \sim \alpha(v) \text{ for all } v \in E_1). \tag{2.16}$$

Corollary 2.2 *The isomorphism class of a clutching (G, ϕ)-vector bundle $E_1 \cup_\alpha E_2$ depends only on the homotopy class of the isomorphism $\alpha : E_1|_A \longrightarrow E_2|_A$.*

Proof Suppose $\alpha_t : E_1|_A \longrightarrow E_2|_A$ is a homotopy of (G, ϕ)-isomorphisms. Then we can construct an (G, ϕ)-isomorphism of (G, ϕ)-vector bundles over $A \times I$ as follows

$$
\begin{aligned}
\alpha : (\pi_1^* E_1)|_{A \times I} &\longrightarrow (\pi_1^* E_2)|_{A \times I} \\
(a, t, v) &\longmapsto (a, t, \alpha_t(v)).
\end{aligned}
\tag{2.17}
$$

where $\pi_1 : X \times I \longrightarrow X$ is the projection onto the first factor. So we can construct the clutching of $\pi_1^* E_1$ and $\pi_1^* E_2$

$$\pi_1^* E_1 \cup_\alpha \pi_1^* E_2 \tag{2.18}$$
$$\downarrow$$
$$X \times I.$$

For $t \in I$, define the function

$$f_t : X \longrightarrow X \times I$$
$$x \longmapsto (x, t). \tag{2.19}$$

Note for $t \in I$ we have an (G, ϕ)-isomorphism of (G, ϕ)-vector bundles over X

$$E_1 \cup_{\alpha_t} E_2 \xrightarrow{\;\cong\;} f_t^*(\pi_1^* E_1 \cup_\alpha \pi_1^* E_2). \tag{2.20}$$

So apply Lemma 2.7 to the (G, ϕ)-vector bundle $\pi_1^* E_1 \cup_\alpha \pi_1^* E_2$ and the (G, ϕ)-homotopic functions f_0, f_1. $\qquad\qquad\qquad\qquad\qquad\qquad\qquad\qquad\qquad\square$

Let (X, A) be a closed G-pair with A G-contractible. If E is a (G, ϕ)-vector bundle over X, then by Corollary 2.1 there exist a (G, ϕ)-isomorphism $\alpha : E|_A \longrightarrow A \times V$ with V a representation of (G, ϕ). We refer to α as a **trivialization** of E over A. Let $\pi_2 : A \times V \longrightarrow V$ be the projection onto the second factor and define an equivalence relation on $E|_A$ by

$$e \sim e' \text{ if and only if } \pi_2(\alpha(e)) = \pi_2(\alpha(e')) \tag{2.21}$$

and extend it by the identity on $E|_{X \setminus A}$. Denote by E/α the quotient of E by this equivalence relation. Then $E/\alpha \longrightarrow X/A$ is a (G, ϕ)-vector bundle, one just needs to verify the local triviality around the point A/A but this is an application of Lemma 2.6. Moreover:

Corollary 2.3 *Let (X, A) be a closed G-pair with A G-contractible, $E \longrightarrow X$ a (G, ϕ)-vector bundle over X and $\alpha : E|_A \longrightarrow A \times V$ a trivialization of E over a closed G-subspace A. Then the isomorphism class of the collapsing construction E/α depends only on the homotopy class of α.*

Proof Similar to the proof of Corollary 2.2, a G-homotopy between trivializations α_0 and α_1 induces an isomorphism $\beta : (E \times I)|_{A \times I} \longrightarrow A \times I \times V$ of (G, ϕ)-vector bundles over $A \times I$. Consider the natural G-map

$$f : (X/A) \times I \longrightarrow (X \times I)/(A \times I)$$
$$([x], t) \longrightarrow [x, t] \tag{2.22}$$

This a G-homotopy between $f_0 = f|_{(X/A)\times\{0\}}$ and $f_1 = f|_{(X/A)\times\{1\}}$. For $i = 0, 1$, we have (G, ϕ)-isomorphisms of (G, ϕ)-vector bundles over $(X/Y) \times I$

$$E/\alpha_i \longrightarrow f^* \left((E \times I)/\beta\right)|_{(X/Y)\times\{i\}}. \tag{2.23}$$

So apply Lemma 2.7 to the (G, ϕ)-vector bundle $(E \times I)/\beta$ and the maps f_0 and f_1. $\qquad\square$

For X, Y two G-spaces, denote by $[X, Y]^G$ the set of G-homotopy classes of G-maps.

Lemma 2.8 *Let A be a G-contractible G-space, V a representation of (G, ϕ) and suppose it decomposes as $V \cong \left(A_1^{a_1} \oplus \ldots \oplus A_m^{a_m}\right) \oplus \left(B_1^{b_1} \oplus \ldots \oplus B_n^{b_n}\right) \oplus \left(C_1^{c_1} \oplus \ldots \oplus C_l^{c_l}\right)$ where A_i, B_j, C_k are non isomorphic irreducible representations of real, complex and quaternionic type, respectively. Then we have a bijection*

$$[A, GL(V)]^G \xrightarrow{\;\cong\;} \Pi_i\{\pm 1\}^{a_i}. \tag{2.24}$$

where the action of G on $GL(V)$ is given by $(g \cdot T)(v) = g \cdot T(g^{-1} \cdot v)$ with $g \in G$, $T \in GL(V)$ and $v \in V$. The set $\Pi_i\{\pm 1\}^{a_i}$ is the discrete space of $(\sum_i a_i)$-tuples in $\{\pm 1\}$.

Proof Suppose $* \in A$ is a fixed point of the group G. A G-homotopy $H : A \times I \longrightarrow A$, between the identity map id_A and a the constant map $A \longrightarrow \{*\}$, gives rise to a bijection

$$[A, \mathrm{GL}(V)]^G \longrightarrow [*, \mathrm{GL}(V)]^G. \tag{2.25}$$

Now we have the equality

$$[*, \mathrm{GL}(V)]^G = \pi_0^G(\mathrm{Iso}_{\mathbf{Rep}(G,\phi)}(V)) \tag{2.26}$$

where $\pi_0^G(X)$ is the set of G-path components of X. Finally, by Schur's lemma in Lemma 1.1 we have the bijection

$$\pi_0^G\left(\mathrm{Iso}_{\mathbf{Rep}(G,\phi)}(V)\right) \cong \pi_0^G\left(\bigoplus_i(\mathbb{R}^*)^{a_i} \oplus \bigoplus_j(\mathbb{C}^*)^{b_j} \oplus \bigoplus_k(\mathbb{H}^*)^{c_k}\right) \tag{2.27}$$

$$\cong \pi_0^G\left(\bigoplus_i(\mathbb{R}^*)^{a_i}\right) \times \pi_0^G\left(\bigoplus_j(\mathbb{C}^*)^{b_j}\right) \times \pi_0^G\left(\bigoplus_k(\mathbb{H}^*)^{c_k}\right) \tag{2.28}$$

$$\cong \Pi_i\{\pm 1\}^{a_i}. \tag{2.29}$$

$\qquad\square$

Remark 2.3 As we have seen, Lemma 2.3 shows that the G-homotopy classes $[A, \mathrm{GL}(V)]^G$ depend only on the irreducible representations of real type in the decomposition of V. However, we always can choose classes in $\mathbb{R}^+ = \{x \in \mathbb{R} \mid x > 0\}$ or in $\mathbb{R}^- = \{x \in \mathbb{R} \mid x < 0\}$ because they always exist, and **by convention we choose the positive part** $\mathbb{R}^+ \subset \mathbb{R}^*$.

An example of this choice, and possibly the most important, is the following: Given a (G, ϕ)-vector bundle E over X, then there exist trivializations $\alpha : E|_A \longrightarrow A \times V$. Two of these trivializations $\alpha_0, \alpha_1 : E_A \longrightarrow A \times V$ define a map $\alpha_1^{-1}\alpha_0 : A \times V \longrightarrow A \times V$ which is of the form $(x, v) \longmapsto (x, \Psi_x(v))$. The map $\Psi : A \longrightarrow \mathrm{GL}(V)$ is G-equivariant, so its class is in $[\Psi] \in [A, \mathrm{GL}(V)^G] \cong \pi_0^G(\mathrm{Iso}_{\mathbf{Rep}(G,\phi)}(V)) = \{\pm 1\}^{\oplus_i a_i}$. We choose α_0, α_1 such that $[\Psi]$ is represented by $(1, 1, \ldots, 1) \in \{\pm 1\}^{\oplus_i a_i}$.

Corollary 2.4 *Let (X, A) be a (G, ϕ)-pair with A (G, ϕ)-contractible. Then the canonical projection $f : X \longrightarrow X/A$ induces a bijection*

$$f^* : \mathbf{Vect}_{(G,\phi)}(X/A)/\cong \longrightarrow \mathbf{Vect}_{(G,\phi)}(X)/\cong . \qquad (2.30)$$

Proof Given a (G, ϕ)-vector bundle E over X, choose a trivialization $\alpha : E|_A \longrightarrow A \times V$ and then consider the (G, ϕ)-vector bundle $E/\alpha \longrightarrow X/A$. So we have map

$$\mathbf{Vect}_{(G,\phi)}(X)/ \longrightarrow \mathbf{Vect}_{(G,\phi)}(X/A)/\cong . \qquad (2.31)$$

Let us see that this map only depends on the isomorphism class of E. Let F be (G, ϕ)-vector bundle over X isomorphic to E, choose a trivialization $\beta : F|A \longrightarrow A \times V$ such that $\beta^{-1}\alpha\pi$ produces a (G, ϕ)-map $A \longrightarrow GL(V)$ represented by a class of the form $(1, 1 \ldots, 1) \in [A, \mathrm{GL}(V)]^G$, following our convention in Remark 2.3. Then by Corollary 2.3 E/α is isomorphic to F/β. $\qquad \square$

Remark 2.4 For equivariant complex vector bundles the quotient map $f : X \longrightarrow X/A$ induces a bijection between the corresponding isomorphism classes of equivariant complex vector bundles over X and X/A, see for example [13, Sect. 2.10] or [14, Sect. 1.3.8].

The usual **suspension** of a G-space X is $SX := C_+(X) \cup C_-(X)$ where $C_-(X) = X \times [0, \frac{1}{2}]/(x, 0) \sim (y, 0)$ and $C_+(X) = X \times [\frac{1}{2}, 1]/(x, 1) \sim (y, 1)$. We put the trivial G action on $[0, 1]$, so $C_-(X)$, $C_+(X)$ and SX are G-spaces. The two cones $C_-(X)$ and $C_+(X)$ are G-contractible.

Denote by $\mathbf{Vect}_{(G,\phi)}^n(SX)$ the set of (G, ϕ)-vector bundles over X of dimension n.

Corollary 2.5 *There is a natural bijection*

$$\mathbf{Vect}_{(G,\phi)}^n(SX)/\cong \xrightarrow{\ \cong\ } \bigcup_{\substack{\textit{Isomorphism classes of n-dimensional} \\ \textit{representations } V \textit{ of } (G, \phi)}} [X, GL(V)]^G \qquad (2.32)$$

Proof If E is a n-dimensional (G, ϕ)-vector bundle over SX, then the restrictions of E to $C_+(X)$ and $C_-(X)$ are trivial, and by Corollary 2.1 we have trivializations $\alpha_\pm : E|_{C_\pm(X)} \longrightarrow C_\pm(X) \times V$. We can restrict the trivializations to $X = X \times \{\frac{1}{2}\}$ and get an isomorphism of (G, ϕ)-vector bundles

$$\alpha_+ \circ \alpha_-^{-1} : X \times V \longrightarrow X \times V \tag{2.33}$$

This map is of the form $(x, e) \longmapsto (x, \psi_x(e))$ where $\psi_x \in \mathrm{GL}(V)$ so we have a map

$$\psi : X \longrightarrow \mathrm{GL}(V). \tag{2.34}$$

This map commutes with the action of (G, ϕ):

$$\alpha_+ \circ \alpha_-^{-1}(g \cdot (x, v)) = g \cdot \alpha_+ \circ \alpha_-^{-1}(x, v) \tag{2.35}$$

$$\alpha_+ \circ \alpha_-^{-1}(g \cdot x, g \cdot v) = g \cdot (x, \psi_x(v)) \tag{2.36}$$

$$(g \cdot x, \psi_{g \cdot x}(g \cdot v)) = (g \cdot x, g \cdot \psi_x(v)) \tag{2.37}$$

so $\psi_{g \cdot x}(g \cdot v) = g \cdot \psi_x(v)$. If we put $w = gv$ we get

$$\psi_{g \cdot x}(w) = g \cdot \psi_x(g^{-1}w) \tag{2.38}$$

$$\psi_{g \cdot x}(w) = (g \cdot \psi_x)(w), \tag{2.39}$$

hence $\psi(g \cdot x) = g \cdot \psi(x)$ i.e. $\psi \in [X, \mathrm{GL}(V)]^G$. So we have a map

$$\mathbf{Vect}^n_{(G,\phi)}(SX) \longrightarrow \bigcup_{\substack{\text{Isomorphism classes of } n\text{-dimensional representation } V \text{ of } (G, \phi)}} [X, \mathrm{GL}(V)]^G. \tag{2.40}$$

We can show this function descends to isomorphism classes $\mathbf{Vect}^n_{(G,\phi)}(SX)/\cong$ as in the proof of Corollary 2.4.

The clutching construction defines a map

$$\bigcup_{\substack{\text{Isomorphism classes of } n\text{-dimensional} \\ \text{representation } V \text{ of } (G, \phi)}} [X, \mathrm{GL}(V)]^G \longrightarrow \mathbf{Vect}^n_{(G,\phi)}(SX)/\cong \tag{2.41}$$

and both maps are inverses of each other. $\square$

We need the existence of invariant Hermitian metrics.

Lemma 2.9 *Let E be a (G, ϕ)-vector bundle on a G-space X. Then there exists an Hermitian metric on E invariant under the action of (G, ϕ).*

Proof Given a finite open cover $\{U\}_i$ of X where E is locally trivial, choose Hermitian metrics on each $E|_U \cong U \times \mathbb{C}^n$ and then add them together with a partition of unity, as in [14, Lemma 1.4.10]. So we have a metric $\langle \cdot, \cdot \rangle$ on E. Now we just average the metric to obtain an equivariant one, namely for $x \in X$ and $v, w \in E_x$

$$\langle v, w \rangle_G := \frac{1}{|G|} \sum_G \mathbb{K}^{\phi(g)} \left(\langle gv, gw \rangle \right). \tag{2.42}$$

The complex conjugation is needed to make the first entry linear and the second entry antilinear. $\qquad\square$

Lemma 2.10 *If $p : E \longrightarrow X$ is a (G, ϕ)-vector bundle, then there exist a representation V of (G, ϕ) and a (G, ϕ)-embedding of (G, ϕ)-vector bundles over X*

$$\begin{array}{ccc}
E & \longrightarrow & X \times V \\
& {}_{p}\searrow \quad \swarrow_{\pi_1} & \\
& X. &
\end{array} \tag{2.43}$$

Proof Let $x \in X$. Notice that the sum

$$\bigoplus_{y \in \mathrm{orb}_G(x)} F_y, \tag{2.44}$$

of the fibers of E over the points in the orbit $\mathrm{orb}_G(x)$, is a representation of (G, ϕ). We have a natural (G, ϕ)-embedding of (G, ϕ)-vector bundles over $\mathrm{orb}_G(x)$

$$\Psi_x : E|_{\mathrm{orb}_G(x)} \longrightarrow \mathrm{orb}_G(x) \times \bigoplus_{y \in \mathrm{orb}_G(x)} F_y. \tag{2.45}$$

By Lemma 2.6 there exist an invariant open neighborhood U_x of $\mathrm{orb}_G(x)$ and an extension of Ψ_{U_x} which is still an embedding

$$\Psi_{U_x} : E|_{U_x} \longrightarrow U_x \times \bigoplus_{y \in \mathrm{orb}_G(x)} F_y. \tag{2.46}$$

We can cover X with a finite collection of trivializing open sets $\{U_\alpha\}$ since X is a compact space. Let $\{f_\alpha\}$ be a partition of unity with $\mathrm{supp}(f_\alpha) \subset U_\alpha$. Denote by V the sum $\oplus_\alpha \left(\oplus_{y \in \mathrm{orb}(\alpha)} F_\alpha \right)$ of representations of (G, ϕ). Define the maps

$$\widetilde{\Psi}_\alpha : E \longrightarrow X \times V \tag{2.47}$$

$$v \longmapsto \begin{cases} f_\alpha(p(x))\Psi_{U_\alpha}(v) & \text{if } p(v) \in U_\alpha \\ 0 & \text{if } p(v) \notin U_\alpha. \end{cases} \tag{2.48}$$

The (G, ϕ)-maps $\widetilde{\Psi}_\alpha$ define an embedding $E \longrightarrow X \times V$. $\qquad\square$

Proposition 2.2 *If E is a (G, ϕ)-vector bundle on a G-space X, then there is a trivial (G, ϕ)-vector bundle V and a (G, ϕ)-vector bundle $E^\perp$ such that $E \oplus E^\perp \cong V$.*

Proof By Lemma 2.10 there exists a (G, ϕ)-embedding $E \longrightarrow X \times V$. By Lemma 2.9 there exists an Hermitian metric on $X \times V$ invariant under the action of (G, ϕ). Define $E^\perp$ as the orthogonal complement of E in $X \times V$. $\qquad\square$

Proposition 2.3 *The set of isomorphism classes of stable (G, ϕ)-vector bundles is an abelian group $\widetilde{\mathbf{K}}_G(X)$. It can be identified naturally with a quotient group of $\mathbf{K}_G(X)$.*

Proof Proposition 2.2 guaranties the existence of inverses in $\widetilde{\mathbf{K}}_G(X)$.

Now we must prove there exists an epimorphism $\mathbf{K}_G(X) \longrightarrow \widetilde{\mathbf{K}}_G(X)$. Take an element of $\mathbf{K}_G(X)$ represented by $E - F$. Choose a (G, ϕ)-vector bundle F' such that $F \oplus F' \cong V$ is trivial. We have

$$E - F = E \oplus F' - F \oplus F' \tag{2.49}$$
$$= H - V \tag{2.50}$$

so any element $E - F \in \mathbf{K}_G(X)$ can be represented by $H - V$. Define the homomorphism

$$\mathbf{K}_G(X) \longrightarrow \widetilde{\mathbf{K}}_G(X)$$
$$H - V \longmapsto H. \tag{2.51}$$

Clearly this is an epimorphism and the kernel is generated by elements of the form $V' - V$ which is isomorphic to $\mathbf{R}(G)$. $\qquad\square$

Theorem 2.1 *If $N \le G_0$ is a normal subgroup which acts freely on X, then the projection $\pi : X \longrightarrow X/N$ induces an isomorphism*

$$\pi^* : \mathbf{K}_{G/N}(X/N) \longrightarrow \mathbf{K}_G(X) \tag{2.52}$$

Proof An inverse to π^* is given by the homomorphism $E \longmapsto E/N$. It is the same as [13, Sect. 2.1]. $\qquad\square$

2.4 Magnetic K-Theory as a Cohomology Theory

In this section we are going to enhance the magnetic equivariant K-theory to a cohomology theory for a fixed magnetic group (G, ϕ).

Definition 2.5 Let X be a compact G-space with a base point x_0. The inclusion $x_0 \longrightarrow X$ induces a restriction homomorphism $\psi : \mathbf{K}_G(X) \longrightarrow \mathbf{K}_G(x_0)$ and we define the **reduced K-theory** of X as

$$\widetilde{\mathbf{K}}_G(X) := \ker \psi. \tag{2.53}$$

The K-theory of a compact G-pair (X, Y) is defined as

$$\mathbf{K}_G(X, Y) := \widetilde{\mathbf{K}}_G(X/Y). \tag{2.54}$$

In case $Y = \emptyset$ we set $X/Y = X_+ := X \sqcup \{*\}$ to get $\mathbf{K}_G(X) = \mathbf{K}_G(X, \emptyset)$.

Remark 2.5 This definition of reduced **K**-theory is naturally isomorphic to with the one given in Proposition 2.3.

Definition 2.6 We define the (p, q)-**suspension groups** of the G-pair (X, Y) as

$$\mathbf{K}_G^{p,q}(X, Y) = \mathbf{K}_G(X \times B^{p,q}, X \times S^{p,q} \cup Y \times B^{p,q}) \tag{2.55}$$

The usual suspension groups $\mathbf{K}_G^{-q}$ are given by

$$\mathbf{K}_G^{-q}(X, Y) := \mathbf{K}_G^{0,q}(X, Y). \tag{2.56}$$

In particular, for a base point $x_0 \in X$, we have $\widetilde{\mathbf{K}}_G^{-q}(X) = \widetilde{\mathbf{K}}_G(\Sigma^q X)$ where $\Sigma X = X \times I/(X \times \{0, 1\} \cup \{x_0\} \times I)$ is the **reduced suspension** of (X, x_0).

Example 2.5 Consider the inclusion $\iota : \mathbb{R}^2 = \mathbb{R}^{0,2} \longrightarrow \mathbb{R}^{2,2} = \mathbb{C}^2$, which produces a map

$$P(\mathbb{R}^2) \longrightarrow P(\mathbb{C}^2) \tag{2.57}$$

and induces a morphism

$$\begin{array}{ccc}
\widetilde{\mathbf{K}}_G(P(\mathbb{C}^2)) & \xrightarrow{\;\iota^*\;} & \widetilde{\mathbf{K}}_G(P(\mathbb{R}^2)) \\
\| & & \| \\
\widetilde{\mathbf{K}}_G(B^{1,1}/S^{1,1}) & \xrightarrow{\;\iota^*\;} & \widetilde{\mathbf{K}}_G(B^{0,1}/S^{0,1}) \\
\| & & \| \\
\mathbf{K}_G^{1,1}(*) & \xrightarrow{\;\iota^*\;} & \mathbf{K}_G^{0,1}(*)
\end{array} \tag{2.58}$$

Consider the Hopf line bundle (dual to the tautological (G, ϕ)-vector bundle)

$$H^* = \left\{ (L, p) \in \mathrm{P}(\mathbb{C}^2) \times \mathbb{C}^2 \mid p \in L \right\} \tag{2.59}$$

over $\mathrm{P}(\mathbb{C}^2)$). This is a (G, ϕ)-vector bundle with the action induced by the conjugation action and the homomorphism $G \xrightarrow{\phi} \mathbb{Z}/2$. If we subtract a trivial line bundle, then the **reduced Hopf bundle**, also called the **Bott element** $[H] - \mathbf{1}$, is in the reduced K-group $\widetilde{\mathbf{K}}_G(\mathrm{P}(\mathbb{C}^2))$ and so $\eta := \iota^*(H - 1)$ is the **reduced Hopf bundle** over $\mathrm{P}(\mathbb{R}^2)$. Of course we have a $\mathbb{Z}/2$-homotopy equivalence $B^{1,1}/S^{1,1} \longrightarrow \mathrm{P}(\mathbb{R}^{2,2})$ given by stereographic projection and the Hopf fibration.

Lemma 2.11 *For (X, Y) a G-pair we have an exact sequence*

$$\mathbf{K}_G(X, Y) \xrightarrow{j^*} \mathbf{K}_G(X) \xrightarrow{i^*} \mathbf{K}_G(Y), \tag{2.60}$$

where $i : Y \longrightarrow X$ and $j : (X, \emptyset) \longrightarrow (X, Y)$ are the inclusions.

Proof The composition $i^* j^*$ is induced by the composition $ji : (Y, \emptyset) \longrightarrow (X, Y)$, which factors through (Y, Y), and so the $i^* j^*$ factors through the zero group $\mathbf{K}_G(Y, Y)$ and thus im $j^* \subset \ker i^*$.

Now, suppose that $\xi \in \ker i^*$. Represent ξ as $E - \mathbf{V}$ where E is a (G, ϕ)-vector bundle over X and $\mathbf{V}$ is a trivial (G, ϕ)-vector bundle. Since $i^*(\xi) = 0$ it follows that $[E|_Y] = [\mathbf{V}]$ in $\mathbf{K}_G(Y)$. This implies that for some trivial (G, ϕ)-vector bundle W we have

$$(E \oplus \mathbf{W})|_Y = \mathbf{V} \oplus \mathbf{W}. \tag{2.61}$$

Thus, we have a trivialization α of $(E \oplus \mathbf{W})|_Y$ and hence we have the (G, ϕ)-vector bundle $(E \oplus \mathbf{W})/\alpha \longrightarrow X/Y$. If we define the element $\eta = (E \oplus \mathbf{W})/\alpha - \mathbf{V} \oplus \mathbf{W} \in \widetilde{\mathbf{K}}_G(X/Y) = \mathbf{K}_G(X, Y)$, then

$$j^*(\eta) = E \oplus \mathbf{W} - \mathbf{V} \oplus \mathbf{W} = E - \mathbf{V} = \xi. \tag{2.62}$$

$\square$

Corollary 2.6 *Let (X, Y) be a G-pair and $y \in Y$ a base point. Then we have an exact sequence*

$$\mathbf{K}_G(X, Y) \xrightarrow{j^*} \widetilde{\mathbf{K}}_G(X) \xrightarrow{i^*} \widetilde{\mathbf{K}}_G(Y), \tag{2.63}$$

where $i : Y \longrightarrow X$ and $j : (X, \emptyset) \longrightarrow (X, Y)$ are the inclusions.

Proof It follows from the next commutative diagram

$$0 \longrightarrow \mathbf{K}_G(\{y\}) = \!\!= \mathbf{K}_G(\{y\}) \qquad\qquad (2.64)$$

$$\mathbf{K}_G(X, Y) \xrightarrow{\ j^* \ } \mathbf{K}_G(X) \xrightarrow{\ i^* \ } \mathbf{K}_G(Y)$$

$$\mathbf{K}_G(X, Y) \xrightarrow{\ j^* \ } \widetilde{\mathbf{K}}_G(X) \xrightarrow{\ i^* \ } \widetilde{\mathbf{K}}_G(Y).$$

$\square$

Lemma 2.12 *For a G-pair (X, Y) we have a long exact sequence*

$$\ldots \longrightarrow \mathbf{K}_G^{-1}(X) \longrightarrow \mathbf{K}_G^{-1}(Y) \longrightarrow \mathbf{K}_G(X, Y) \longrightarrow \mathbf{K}_G(X) \longrightarrow \mathbf{K}_G(Y) \quad (2.65)$$

Proof It is enough to show that we have a five term exact sequence

$$\widetilde{\mathbf{K}}_G^{-1}(X) \longrightarrow \widetilde{\mathbf{K}}_G^{-1}(Y) \longrightarrow \mathbf{K}_G(X, Y) \longrightarrow \widetilde{\mathbf{K}}_G(X) \longrightarrow \widetilde{\mathbf{K}}_G(Y). \qquad (2.66)$$

In fact, if this has been established, then replacing (X, Y) by $(S^n X, S^n Y)$ for $n = 1, 2, 3, \ldots$ we obtain a family of exact sequences that can be concatenated to produce a long exact sequence. Then replacing (X, Y) by (X_+, Y_+) we get the long exact sequence of the statement. For a base point $x_0 \in X$, denote by $\widetilde{C}(X) = X \times I /(X \times \{0\} \cup \{x_0\} \times I)$ the reduced cone. So consider the following commutative diagram of G-spaces

$$(2.67)$$

By Corollary 2.6, when we apply $\widetilde{\mathbf{K}}_G(\_)$, the first three rows become exact sequences. All spaces, maps and homotopies involved here are in the category of G-spaces, and the cones are G-contractible. So the same arguments in [13, Sect. 2.6] apply to this context, but of course with the appropriate results about the homotopy properties of (G, ϕ)-vector bundles, G-maps and G-homotopies. $\square$

If X is a locally compact G-space, then the one-point compactification $X \cup \{\infty\}$ is still a G-space with the new point ∞ being a fixed point. This allows us to define the $\mathbf{K}_G$-theory of locally compact G-spaces X as

$$\mathbf{K}_G(X) := \ker\left(\mathbf{K}_G(X \cup \{\infty\}) \longrightarrow \mathbf{K}_G(\infty)\right). \tag{2.68}$$

If (X, Y) is a compact G-pair, then we have a natural G-homeomorphism

$$X/Y \longrightarrow (X\backslash Y) \cup \{\infty\} \tag{2.69}$$

so we have an isomorphism

$$\mathbf{K}_G(X, Y) \cong \mathbf{K}_G(X/Y). \tag{2.70}$$

Corollary 2.7 *For a locally compact G-pair (X, Y) we have a long exact sequence*

$$\ldots \longrightarrow \mathbf{K}_G^{-1}(X) \longrightarrow \mathbf{K}_G^{-1}(Y) \longrightarrow \mathbf{K}_G(X, Y) \longrightarrow \mathbf{K}_G(X) \longrightarrow \mathbf{K}_G(Y) \quad (2.71)$$

Proof The same as 2.4.7 in [18], we have a commutative diagram

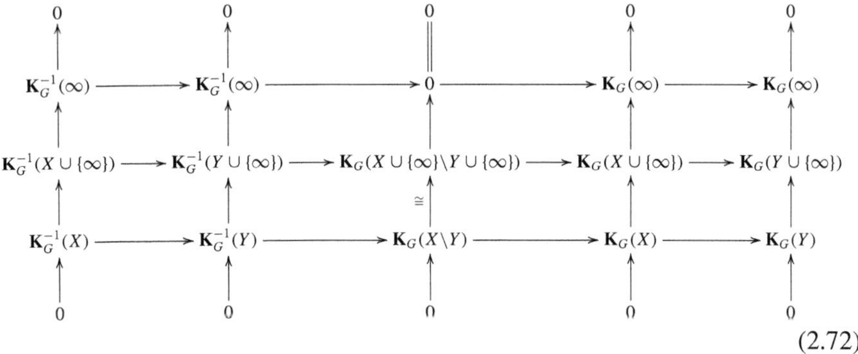

$$\tag{2.72}$$

where the middle row is exact Lemma 2.12 and the columns are just the definition of $\mathbf{K}$-theory for locally compact spaces. $\qquad\square$

Corollary 2.8 *Given a locally compact G-triple (X, Y, Z) there is a long exact sequence*

$$\ldots \longrightarrow \mathbf{K}_G^{-1}(X, Z) \longrightarrow \mathbf{K}_G^{-1}(Y, Z) \longrightarrow \mathbf{K}_G(X, Y) \longrightarrow \mathbf{K}_G(X, Z) \longrightarrow \mathbf{K}_G(Y, Z) \tag{2.73}$$

Proof As in [18, 2.4.14], apply Corollary 2.7 to the G-pair $(X\backslash Z, Y\backslash Z)$. $\qquad\square$

Corollary 2.9 *Given a locally compact G-triple (X, Y, Z) there is a long exact sequence*

$$\ldots \longrightarrow \mathbf{K}_G^{p,1}(X, Z) \longrightarrow \mathbf{K}_G^{p,1}(Y, Z) \longrightarrow \mathbf{K}_G^{p,0}(X, Y) \longrightarrow \mathbf{K}_G^{p,0}(X, Z) \longrightarrow \mathbf{K}_G^{p,0}(Y, Z) \tag{2.74}$$

for each integer $p \geq 0$.

Proof As in [1, p. 373], apply Corollary 2.8 to the G-triple $(X \times B^{p,0}, X \times S^{p,0} \cup Y \times B^{p,0}, X \times S^{p,0})$. $\qquad\square$

2.5 Products and Bott Periodicity

Let E be a (G, ϕ)-vector bundle over a G-space X. Recall that $P(E)$ the **projective bundle** of E, which fiber at $x \in X$ is given by the projective space $P(E_x)$, is also a G-space. The tautological line-bundle H^* is given by

$$H^* = \{(l, v) \in P(E) \times E \mid l \in P(E) \text{ and } v \in l\} \qquad (2.75)$$

$$\pi_1 \downarrow$$

$$P(E),$$

it is naturally a (G, ϕ)-vector bundle. Denote by H its dual line bundle, known as the standard line (G, ϕ)-vector bundle over $P(E)$.

The projection map $P(E) \longrightarrow X$ induces a ring homomorphism $\mathbf{K}_G(X) \longrightarrow \mathbf{K}_G(P(E))$ so that $\mathbf{K}_G(P(E))$ becomes a $\mathbf{K}_G(X)$-algebra. For a particular (G, ϕ)-vector bundle E we can determine the structure of the algebra:

Theorem 2.2 (Bott periodicity) *Let L be a line (G, ϕ)-vector bundle over a G-space X, H the standard line (G, ϕ)-vector bundle over the G-space $P(L \oplus 1)$. Then, as a $\mathbf{K}_G(X)$-algebra, $\mathbf{K}_G(P(L \oplus 1))$ is generated by $[H]$, subject to the single relation*

$$([H] - [1])\,([L][H] - [1]) = 0. \qquad (2.76)$$

Proof As Atiyah points out in [1, Theorem 2.1], we can follow the proof given in [19] and modify the necessary parts. That proof is divided into four parts, we will state the results and sketch the proof in the appropriate equivariant way. To simplify notation set $P := P(L \oplus 1)$.

Part 1. (G, ϕ)-vector bundles over P as clutching bundles via Laurent polynomials

By Lemma 2.9, choose a G-invariant metric in L and let $S \subset L$ be the unit circle bundle in this metric. We identify L with a subspace of P and we can write

$$P = P^0 \cup P^\infty, \qquad S = P^0 \cap P^\infty, \qquad (2.77)$$

where P^0 is the closed disc (G, ϕ)-bundle interior to S and P^∞ is the closed disc (G, ϕ)-bundle exterior to S. So we have the (G, ϕ)-bundles

$$(2.78)$$

Suppose now that E^0, E^∞ are two (G, ϕ)-vector bundles over X and $f : \pi^* E^0 \longrightarrow \pi^* E^\infty$ is an isomorphism of (G, ϕ)-vector bundles over S. Then we can form the (G, ϕ)-vector bundle

$$\pi_0^* E^0 \cup_f \pi_\infty^* E^\infty \tag{2.79}$$

$$\downarrow$$

$$P.$$

We shall denote this (G, ϕ)-vector bundle for brevity by (E^0, f, E^∞) and we shall say that f is a clutching function for (E^0, E^∞).

Lemma 2.13 *Let E be any (G, ϕ)-vector bundle over P and let E^0, E^∞ be the (G, ϕ)-vector bundles over X induced by the 0-section and the ∞-section respectively. Then there exists a (G, ϕ)-isomorphism $f : \pi^* E^0 \longrightarrow \pi^* E^\infty$ such that*

$$E \cong (E^0, f, E^\infty), \tag{2.80}$$

the isomorphism being the natural one on the 0-section and the ∞-section.

Proof Let $s_0 : X \longrightarrow P^0$ be the 0-section. Then $s_0 \circ \pi_0([v : 1]) = [0, 1]$ is G-homotopic to the identity, for example consider the G-homotopy $H_s([v : 1]) = [sv : 1]$. So there exist an (G, ϕ)-isomorphism

$$f_0 : E|_{P^0} \longrightarrow \pi_0^* E^0 = \pi_0^* s_0^* (E|_{P^0}).$$

We can use a similar procedure to obtain an (G, ϕ)-isomorphism

$$f_\infty : E|_{P^\infty} \longrightarrow \pi_\infty^* E^\infty = \pi_\infty^* s_\infty^* (E|_{P^\infty}).$$

If we define $f = f_\infty f_0^{-1}$, then we are done. $\qquad\qquad\qquad\square$

Now we will focus on the clutching functions. The natural inclusion of (G, ϕ)-bundles $\iota : S \longrightarrow L$ defines a G-section $z \in \Gamma_{(G,\phi)}(\pi^* L)$ of $\pi^* L$ given by $z(v) := (v, \iota(v))$. This is depicted as:

$$\begin{array}{ccc}
\pi^* L & \longrightarrow & L \\
\nearrow \Big\downarrow & & \Big\downarrow \pi \\
S & \xrightarrow{\ \pi\ } & X.
\end{array} \tag{2.81}$$

Using the canonical isomorphisms

$$\pi^*(L) \cong \pi^* \mathrm{Hom}(\mathbf{1}, L) \tag{2.82}$$

we may also regard z as a G-section of the (G, ϕ)-vector bundle $\pi^*\mathrm{Hom}(\mathbf{1}, L)$ and, as such, it has an inverse z^{-1} which is a G-section of $\pi^*\mathrm{Hom}(L, \mathbf{1}) \cong \pi^*(L^{-1})$. More generally, for any integer k, we may regard z^k as a G-section of π^*L^k. If $a_k \in \Gamma_{(G,\phi)}(L^{-k})$ then

$$a_k z^k := \pi^*(a_k) \otimes z^k \in \Gamma_{(G,\phi)}(\pi^*\mathbf{1}), \tag{2.83}$$

i.e. it is a function on S. Finally suppose that E^0, E^∞ are two (G, ϕ)-vector bundles on X and that

$$a_k \in \Gamma_{(G,\phi)}\mathrm{Hom}(L^k \otimes E^0, E^\infty), \tag{2.84}$$

then

$$a_k z^k := \pi^*(a_k) \otimes z^k \in \Gamma_{(G,\phi)}\mathrm{Hom}(\pi^*E^0, \pi^*E^\infty). \tag{2.85}$$

If $f = \sum_{-n}^{n} a_k z^k : \pi^*E^0 \longrightarrow \pi^*E^\infty$ is an (G, ϕ)-isomorphism, then it defines a clutching function and we call this a **Laurent clutching function** for (E^0, E^∞). One can prove that there is a (G, ϕ)-isomorphism

$$H^{-k} \cong (\mathbf{1}, z^{-k}, L^{-k}). \tag{2.86}$$

Suppose now that $f \in \Gamma_{(G,\phi)}\mathrm{Hom}(\pi^*E^0, \pi^*E^\infty)$ is any G-section, then we can define its **Fourier coefficients**

$$a_k(x) = \frac{1}{2\pi i} \int_{S_x} f_x z_x^{-k-1} dz_x \tag{2.87}$$

where f_x and z_x denote the restrictions of f, z to S_x and dz_x is therefore a differential on S_x. The section a_k is G-equivariant:

$$(g \cdot a_k)(x) = g a_k(g^{-1}x) \tag{2.88}$$

$$= g \left(\frac{1}{2\pi i} \int_{S_{g^{-1}x}} f_{g^{-1}x} z_{g^{-1}x}^{-k-1} dz_{g^{-1}x} \right) \tag{2.89}$$

$$= \frac{(-1)^{\phi(g)}}{2\pi i} \int_{S_{g^{-1}x}} (g f_{g^{-1}x})(g z_{g^{-1}x})^{-k-1} (g dz_{g^{-1}x}) \tag{2.90}$$

$$= \frac{1}{2\pi i} \int_{S_x} f_x z_x^{-k-1} dz_x \tag{2.91}$$

$$= a_k(x). \tag{2.92}$$

Let s_n be the partial sum

$$s_n = \sum_{-n}^{n} a_k z^k \tag{2.93}$$

and define the **Cesaro means**

$$f_n = \frac{1}{n} \sum_{k=1}^{n} s_k. \tag{2.94}$$

Lemma 2.14 *Let f be any clutching function for (E^0, E^∞), f_n the sequence of Cesaro means of the Fourier series of f. Then f_n converges uniformly to f and hence is a Laurent clutching function for all sufficiently large n.*

Proof This is a extension of the Fejér's theorem, see Chap. 2 of [20] for a proof. $\square$

Part 2. Linearization of Laurent clutching functions

All the isomorphisms presented here are (G, ϕ)-isomorphisms of (G, ϕ)-vector bundles, since they are given in terms of z^k and a_k which are G-sections.

Part 3. Linear clutching functions

Let p be a linear clutching function for (E^0, E^∞) and let the endomorphisms $Q^0 : E^0 \longrightarrow E^0$ and $Q^\infty : E^\infty \longrightarrow E^\infty$ be given by

$$Q_x^0 = \frac{1}{2\pi i} \int_{S_x} p_x^{-1} dp_x, \qquad Q_x^\infty = \frac{1}{2\pi i} \int_{S_x} dp_x\, p_x^{-1}. \tag{2.95}$$

These are (G, ϕ)-endomorphisms, in fact for $g \in G$

$$(g \cdot Q^0)_x = g Q_{gx}^0 \tag{2.96}$$

$$= g \left(\frac{1}{2\pi i} \int_{S_{gx}} p_{gx}^{-1} dp_{gx} \right) \tag{2.97}$$

$$= \frac{(-1)^{\phi(g)}}{2\pi i} \int_{S_{gx}} g p_{gx}^{-1} g\, dp_{gx} \tag{2.98}$$

$$= \frac{1}{2\pi i} \int_{S_x} p_x^{-1} dp_x, \tag{2.99}$$

and similarly for Q^∞. Again, all the results in this part follow quite formally.

Part 4. Construction of an inverse.

Consider the homomorphism

$$\mu : \mathbf{K}_G(X)[t]/(t - 1)([L]t - 1) \longrightarrow \mathbf{K}_G(P)$$
$$t \longmapsto [H]. \tag{2.100}$$

Then using the previous parts one can construct an inverse for μ. As we mention before, the same lines and the rest of the proof in [19] apply quite formally. $\square$

Corollary 2.10 *We have an isomorphism:*

$$\mathbf{K}_G(\mathrm{P}(\mathbb{C}^2)) \cong \frac{\mathbf{R}(G)[H]}{([H] - [\mathbf{1}])^2}. \tag{2.101}$$

Proof Apply Theorem 2.2 to the space $X = *$ and $L = \mathbf{1}$. In this case $\mathrm{P}(L \oplus \mathbf{1}) \cong \mathrm{P}(\mathbb{C}^2)$ and by Example 2.4 we know $\mathbf{K}_G(*) = \mathbf{R}(G)$ is the representation ring of (G, ϕ). $\qquad\qquad\square$

Corollary 2.11 *The* $\mathbf{R}(G)$*-module* $\widetilde{\mathbf{K}}_G(\mathrm{P}(\mathbb{C}^2))$ *is free with generator* $[H] - [\mathbf{1}]$.

In order to present the most classical version of the periodicity theorem we need to talk about external products.

Definition 2.7 Let $E \longrightarrow X$ and $F \longrightarrow Y$ be (G, ϕ)-vector bundles over G-spaces X and Y. The **external tensor product** of E and F over $X \times Y$ is defined as:

$$E \boxtimes F := \pi_1^*(E) \otimes \pi_2^*(F) \tag{2.102}$$

The map $(E, F) \longmapsto E \boxtimes F$ induces a pairing

$$\mathbf{K}_G(X) \otimes \mathbf{K}_G(Y) \longrightarrow \mathbf{K}_G(X \times Y)$$
$$\alpha \otimes \beta \longmapsto \alpha * \beta \tag{2.103}$$

which is called **external product**.

If $\Delta : X \longrightarrow X \times X$ denotes the diagonal map, then the composition

$$\mathbf{K}_G(X) \otimes \mathbf{K}_G(X) \longrightarrow \mathbf{K}_G(X \times X) \xrightarrow{\Delta^*} \mathbf{K}_G(X) \tag{2.104}$$

is called the **internal product** and we will denoted it by $\alpha\beta = \Delta^*(\alpha * \beta)$. We need also the reduced external product and to this end we need the following classical results.

Corollary 2.12 *If (X, Y) is a G-pair with Y a G-retract of X, then for all $n \geq 0$, the sequence*

$$0 \longrightarrow \mathbf{K}_G^{-n}(X, Y) \longrightarrow \mathbf{K}_G^{-n}(X) \longrightarrow \mathbf{K}_G^{-n}(Y) \longrightarrow 0 \tag{2.105}$$

is a split short exact sequence, so $\mathbf{K}_G^{-n}(X) \cong \mathbf{K}_G^{-n}(X, Y) \oplus \mathbf{K}_G^{-n}(Y)$.

If (X, Y) is a G-pair of based spaces, the projection maps $\pi_1 : X \times Y \longrightarrow X$, $\pi_2 : X \times Y \longrightarrow Y$ induce an isomorphism for all $n \geq 0$

$$\widetilde{\mathbf{K}}_G^{-n}(X \times Y) \cong \widetilde{\mathbf{K}}_G^{-n}(X \wedge Y) \oplus \widetilde{\mathbf{K}}_G^{-n}(X) \oplus \widetilde{\mathbf{K}}_G^{-n}(Y) \tag{2.106}$$

Proof For the first part we see that the long exact sequence of Lemma 2.12 splits using the retract.

For the second observe that X is a G-retract of $X \times Y$, and Y is a G-retract of $X \times Y/X$. The result follows by applying twice the first result. $\square$

By the definition of reduced **K**-theory we have the following isomorphism

$$\mathbf{K}_G(X) \otimes \mathbf{K}_G(Y) \cong \left(\widetilde{\mathbf{K}}_G(X) \oplus \mathbf{K}_G(*)\right) \otimes \left(\widetilde{\mathbf{K}}_G(X) \oplus \mathbf{K}_G(*)\right) \tag{2.107}$$

$$\cong \left(\widetilde{\mathbf{K}}_G(X) \otimes \widetilde{\mathbf{K}}_G(Y)\right) \oplus \left(\widetilde{\mathbf{K}}_G(X) \otimes \mathbf{K}_G(*)\right) \oplus \left(\mathbf{K}_G(*) \otimes \widetilde{\mathbf{K}}_G(Y)\right) \tag{2.108}$$

$$\oplus \left(\mathbf{K}_G(*) \otimes \mathbf{K}_G(*)\right) \tag{2.109}$$

Using the splitting and the Lemma 2.12 we can see the external tensor product $\mathbf{K}_G(X) \otimes \mathbf{K}_G(Y) \longrightarrow \mathbf{K}_G(X \times Y)$ restrict to four maps:

$$\widetilde{\mathbf{K}}_G(X) \otimes \widetilde{\mathbf{K}}_G(Y) \longrightarrow \widetilde{\mathbf{K}}_G(X \wedge Y), \tag{2.110}$$

$$\widetilde{\mathbf{K}}_G(X) \otimes \mathbf{K}_G(*) \longrightarrow \widetilde{\mathbf{K}}_G(X), \tag{2.111}$$

$$\mathbf{K}_G(*) \otimes \widetilde{\mathbf{K}}_G(Y) \longrightarrow \widetilde{\mathbf{K}}_G(Y), \tag{2.112}$$

$$\mathbf{K}_G(*) \otimes \mathbf{K}_G(*) \longrightarrow \mathbf{K}_G(*). \tag{2.113}$$

The first map $\widetilde{\mathbf{K}}_G(X) \otimes \widetilde{\mathbf{K}}_G(Y) \longrightarrow \widetilde{\mathbf{K}}_G(X \wedge Y)$ is called **reduced external tensor product**, the following two maps induce a structure of $\mathbf{K}_G(*) = \mathbf{R}(G)$-module on reduced $\mathbf{K}_G$-theory. Using this reduced tensor product and the natural homeomorphism

$$B^{p,q} \times B^{s,t} \cong B^{p+s,q+t} \tag{2.114}$$

we have a natural product

$$\mathbf{K}_G^{p,q}(X, Y) \otimes \mathbf{K}_G^{s,t}(X', Y') \longrightarrow \mathbf{K}_G^{p+s,q+t}(X \times X', X \times Y' \cup Y \times X'). \tag{2.115}$$

Recall the Bott element given in Example 2.5

$$b = [H] - \mathbf{1} \in \mathbf{K}_G^{1,1}(*) = \mathbf{K}_G(B^{1,1}, S^{1,1}) = \widetilde{\mathbf{K}}_G(\mathrm{P}(\mathbb{C}^2)). \tag{2.116}$$

The **Bott homomorphism** is given by

$$\beta : \mathbf{K}_G^{p,q}(X, Y) \longrightarrow \mathbf{K}_G^{p+1,q+1}(X, Y)$$

$$a \longmapsto a * b. \tag{2.117}$$

Theorem 2.3 *((1, 1)-periodicity) The Bott homomorphism* $\beta : \mathbf{K}_G^{p,q}(X, Y) \longrightarrow \mathbf{K}_G^{p+1,q+1}(X, Y)$ *is an isomorphism.*

Proof We can assume $p = q = 0$ replacing (X, Y) by $(X \times B^{p,q}, X \times S^{p,q} \cup Y \times B^{p,q})$. Also, we can assume Y to be a point, since $\mathbf{K}_G(X, Y) = \widetilde{\mathbf{K}}_G(X/Y) = \widetilde{\mathbf{K}}_G(X/Y, \{Y\})$.

In this situation $\mathbf{K}_G(X, *) = \widetilde{\mathbf{K}}_G(X)$, $\mathbf{K}_G^{1,1}(X, *) = \mathbf{K}_G(X \wedge \mathrm{P}(\mathbb{C}^2))$ and β can be written as the composition

$$\widetilde{\mathbf{K}}_G(X) \xrightarrow{\otimes b} \widetilde{\mathbf{K}}_G(X) \otimes_{R(G)} \widetilde{\mathbf{K}}_G(\mathrm{P}(\mathbb{C}^2)) \xrightarrow{\quad * \quad} \widetilde{\mathbf{K}}_G(X \wedge \mathrm{P}(\mathbb{C}^2)). \tag{2.118}$$

By Corollary 2.11 the first map is an isomorphism. So we need to show the second map is an isomorphism. By the previous morphism relating the external product and the reduced external product we just have to prove the map

$$\mathbf{K}_G(X) \otimes_{\mathbf{R}(G)} \mathbf{K}_G(\mathrm{P}(\mathbb{C}^2)) \longrightarrow \mathbf{K}_G(X \times \mathrm{P}(\mathbb{C}^2)) \tag{2.119}$$

is an isomorphism. But by Theorem 2.2 we have

$$\mathbf{K}_G(\mathrm{P}(\mathbb{C}^2)) \cong \frac{\mathbf{R}(G, \phi)[H]}{([H] - [\mathbf{1}])^2} \tag{2.120}$$

and of course there is a natural isomorphism

$$\mathbf{K}_G(X) \otimes_{\mathbf{R}(G)} \frac{\mathbf{R}(G)[H]}{([H] - [\mathbf{1}])^2} \cong \frac{\mathbf{K}_G(X)[H]}{([H] - [\mathbf{1}])^2} \tag{2.121}$$

which commutes with the internal tensor product. $\qquad\qquad\square$

Definition 2.8 Let the positive degree magnetic equivariant K-theory groups be:

$$\mathbf{K}_G^p(X, Y) := \mathbf{K}_G^{p,0}(X, Y) \qquad\qquad \text{for } p > 0. \tag{2.122}$$

Corollary 2.13 *There is a natural isomorphism*

$$\mathbf{K}_G^{p,q} \cong \mathbf{K}_G^{p-q} \tag{2.123}$$

for all $p, q \geq 0$.

Proof If $p < q$, apply p times Theorem 2.3 and then the definition of the suspension of Eq. (2.56). If $p \geq q$, apply q times Theorem 2.3. $\qquad\qquad\square$

2.6   Clifford Modules

In this section we introduce bigraded Clifford modules in the context of magnetic groups. These are used in Sect. 2.7 to compute the K-theory of spheres with or without involution.

Let $\mathrm{Cliff}(\mathbb{R}^{p,q})$ denote the Clifford algebra on $\mathbb{R}^{p,q}$ with the quadratic form

$$-\left(\sum_{j=1}^{q} x_j^2 + \sum_{i=1}^{p} y_i^2\right) \tag{2.124}$$

A more explicit description is the following

$$\mathrm{Cliff}(\mathbb{R}^{p,q}) = \left\langle 1, e_1, e_2, \ldots, e_q, e_{q+1}, e_{q+2}, \ldots, e_{p+q} \;\middle|\; \begin{matrix} e_j^2 = -1 \\ e_j e_k = -e_k e_j \end{matrix} \right\rangle \tag{2.125}$$

This algebra has a natural action of (G, ϕ) given by

$$g \cdot e_j = \begin{cases} e_j & \text{if } 1 \leq j \leq q \\ (-1)^{\phi(g)} e_j & \text{if } q+1 \leq j \leq q+p. \end{cases} \tag{2.126}$$

Remark 2.6 There is another natural Clifford algebra over $\mathbb{R}^{q+p} = \mathbb{R}^q \oplus \mathbb{R}^p$, which is $\mathbb{R}^{p,q}$ as vector spaces over the real numbers) induced by the quadratic form

$$-\sum_{i}^{q} x^2 + \sum_{j}^{p} y^2. \tag{2.127}$$

Denote this Clifford algebra by $C_{p,q}$. More explicitly we have

$$C_{p,q} = \left\langle 1, e_1, e_2, \ldots, e_q, e_{q+1}, e_{q+2}, \ldots, e_{p+q} \;\middle|\; \begin{matrix} e_j^2 = -1 & 1 \leq j \leq q \\ e_j^2 = 1 & q+1 \leq j \leq q+p \\ e_j e_k = -e_k e_j & \end{matrix} \right\rangle. \tag{2.128}$$

Endow $C_{p,q}$ with the *trivial involution* and hence with the trivial action of G. Later we will compare the categories of Clifford modules over $\mathrm{Cliff}(\mathbb{R}^{p,q})$ and $C_{p,q}$. Denote by C_q the Clifford algebra $C_{0,q}$.

Definition 2.9 A **magnetic** $\mathrm{Cliff}(\mathbb{R}^{p,q})[G, \phi]$-**module** is a $\mathbb{Z}_2$-graded complex vector space $M = M_0 \oplus M_1$ together with:

- a graded $\mathbb{C}$-linear action of $\mathrm{Cliff}(\mathbb{R}^{p,q})$,
- a representation of the magnetic point group (G, ϕ) of degree zero such that

$$g \cdot (zm) = (g \cdot z)(g \cdot m) \tag{2.129}$$

for $g \in G$, $z \in \mathrm{Cliff}(\mathbb{R}^{p,q})$ and $m \in M$.

Denote by $M^{p,q}(G, \phi)$ the Grothendieck group of isomorphism classes of $\mathrm{Cliff}(\mathbb{R}^{p,q})[G, \phi]$-modules.

If instead we choose the Clifford algebra $C_{p,q}$, we have:

Definition 2.10 The magnetic $C_{p,q}[G, \phi]$-modules are $\mathbb{Z}_2$-graded complex vector spaces $M = M_0 \oplus M_1$ together with:

- a graded $\mathbb{C}$-linear action of $C_{p,q}$,
- a representation of the magnetic point group (G, ϕ) of degree zero such that

$$g \cdot (zm) = z(g \cdot m) \tag{2.130}$$

for $g \in G$, $z \in C_{p,q}$ and $m \in M$.

Denote by $\widehat{M}^{p,q}(G, \phi)$ the Grothendieck group of the isomorphism classes of $C_{p,q}[G, \phi]$-modules.

The difference with respect to the $\mathrm{Cliff}(\mathbb{R}^{p,q})[G, \phi]$-modules is that the action of G is trivial on $C_{p,q}$ and non-trivial on $\mathrm{Cliff}(\mathbb{R}^{p,q\cdot})$.

Remark 2.7 We will use both groups $M^{p,q}(G, \phi)$ and $\widehat{M}^{p,q}(G, \phi)$, the first one to produce a well-defined morphism to the K-theory groups, and the second one to perform calculations of the coefficients.

Example 2.6 Let $(G, \phi) = (\mathbb{Z}/2, \mathrm{id})$ and M be a $\mathrm{Cliff}(\mathbb{R}^{p,q})[G, \phi]$-module. Then

- the non-trivial element of $G = \mathbb{Z}/2$ acts on M in an antilinear way; let us denote such action by $m \longmapsto \overline{m}$,
- the Clifford algebra $\mathrm{Cliff}(\mathbb{R}^{p,q})$ acts on M and these two actions interact as follows

$$\overline{zm} = 1 \cdot (zm) \tag{2.131}$$
$$= 1 \cdot (z)1 \cdot (m) \tag{2.132}$$
$$= \overline{z} \cdot \overline{m} \tag{2.133}$$

where $\overline{z} := 1 \cdot z$ denotes the action of $\mathbb{Z}/2$ on $\mathrm{Cliff}(\mathbb{R}^{p,q})$.

So M is a real $\mathbb{Z}/2$-graded $\mathrm{Cliff}(\mathbb{R}^{p,q})$-module in the sense of [1].

Example 2.7 Let $(G, \phi) = (G_0 \rtimes \mathbb{Z}/2, \pi_2)$ and M a magnetic $\mathrm{Cliff}(\mathbb{R}^{0,q})[G, \phi]$-module. Then M is a $\mathbb{Z}/2$-graded complex vector space together with:

- a $\mathbb{C}$-linear graded action of $\mathrm{Cliff}(\mathbb{R}^{0,q}) = C_{0,q} = C_q$,
- an antilinear action given by the element $(e, 1) \in G_0 \rtimes \mathbb{Z}/2$ of degree zero, denoted by $m \longmapsto \overline{m}$. This map commutes with the action of the Clifford algebra because

$$\overline{zm} = (e, 1) \cdot (zm) \tag{2.134}$$
$$= ((e, 1) \cdot z)((e, 1) \cdot m) \tag{2.135}$$
$$= z \cdot \overline{m} \qquad (\mathrm{Cliff}(\mathbb{R}^{0,q}) \text{ has the trivial involution}) \tag{2.136}$$

for $z \in \mathrm{Cliff}(\mathbb{R}^{0,q})$ and $m \in M$.

- The elements of the form $(g, 0) \in G_0 \rtimes \mathbb{Z}/2$ act linearly on M, $m \mapsto g \cdot m$. This action is of degree zero, commutes with the action of C_q, and is such that

$$\begin{aligned}
\overline{g \cdot m} &= (e, 1)(g, 0) \cdot m && (2.137) \\
&= (e, 1)(g, 0)(e, 1)^{-1}(e, 1) \cdot m && (2.138) \\
&= (1 \cdot g, 1)(e, 1) \cdot m && (2.139) \\
&= (\overline{g}, 0) \cdot m && (2.140) \\
&= \overline{g} \cdot \overline{m} && (2.141)
\end{aligned}$$

where $\overline{g} := 1 \cdot g$ denotes the action of $\mathbb{Z}/2$ on G_0.

So M is a real graded $C_q[G_0]$-module in the sense of [2].

As in the classical (complex) case [21], the Real case [1], or the Real equivariant case [2], there is an **Atiyah-Bott-Shapiro homomorphism**

$$\begin{aligned}
\alpha : M^{p,q}(G, \phi) &\longrightarrow \mathbf{K}_G(D^{p,q}, S^{p,q}) \\
M_0 \oplus M_1 &\longmapsto (D^{p,q} \times M_0, D^{p,q} \times M_1, \sigma) && (2.142)
\end{aligned}$$

where $\sigma : S^{p,q} \times M_0 \longrightarrow S^{p,q} \times M_1$ is given by $\sigma(z, m) = (z, zm)$. Of course, the space $D^{p,q} \times M_i$ is a G-space and this action commutes with the projection onto the first coordinate

$$\pi_1 : D^{p,q} \times M_i \longrightarrow D^{p,q}. \tag{2.143}$$

So the vector bundles $D^{p,q} \times M_i \xrightarrow{\pi_1} D^{p,q}$ are (G, ϕ)-vector bundles and the map σ is a morphism of (G, ϕ)-vector bundles. Here we are using the alternative definition of the K-theory groups presented in Definition 2.14 through sequences of complexes of vector bundles.

There is a restriction morphism

$$\mathrm{res} : M^{p,q+1}(G, \phi) \longrightarrow M^{p,q}(G, \phi) \tag{2.144}$$

induced by the inclusion $\mathbb{R}^{p,q} \subset \mathbb{R}^{p,q+1}$. So we have a sequence

$$M^{p,q+1}(G, \phi) \xrightarrow{\mathrm{res}} M^{p,q}(G, \phi) \xrightarrow{\alpha} \mathbf{K}_G(D^{p,q}, S^{p,q}) \longrightarrow 0 \tag{2.145}$$

which will be proven to be exact in Proposition 2.5.

Lemma 2.15 *The composition $\alpha \circ \mathrm{res}$ is zero.*

Proof If $M = M_0 \oplus M_1$ is a $\mathrm{Cliff}(\mathbb{R}^{p,q+1})[G, \phi]$ module, then as said before we have a morphism

$$\sigma : S^{p,q+1} \times M_0 \longrightarrow S^{p,q+1} \times M_1. \tag{2.146}$$

We have the decomposition $S^{p,q+1} = S_+^{p,q_1} \cup S_-^{p,q+1}$ where $S_\pm^{p,q+1} = \{(x_1, \ldots, x_{q+1},$ $y_1, \ldots, y_p) \in S^{p,q+1} \mid \pm x_{q+1} \geq 0\}$ and this decomposition respects the involution. Notice we have a homeomorphism

$$\Psi : S_+^{p,q+1} \xrightarrow{\;\cong\;} D^{p,q}$$
$$(x_1, \ldots, x_{q+1}, y_1, \ldots, y_p) \longmapsto (x_1, \ldots, x_q, y_1, \ldots, y_p). \tag{2.147}$$

Clearly Ψ is a map which commutes with the involution in both spaces. So we get an isomorphism of G-vector bundles

$$D^{p,q} \times M_0 \xrightarrow{(\Psi \times \mathrm{id})\circ\sigma\circ(\Psi^{-1}\times \mathrm{id})} D^{p,q} \times M_1 \tag{2.148}$$

$$\pi_1 \searrow \qquad \swarrow \pi_1$$
$$D^{p,q}$$

which restricts to the isomorphism $\sigma : S^{p,q} \times M_0 \longrightarrow S^{p,q} \times M_1$. $\qquad\qquad\square$

So we have an Atiyah-Bott-Shapiro homomorphism

$$M^{p,q}(G, \phi)/\mathfrak{res}(M^{p,q+1}(G, \phi)) \xrightarrow{\;\alpha\;} \mathbf{K}_G(D^{p,q}, S^{p,q}) . \tag{2.149}$$

Now it is time to compare the two types of Clifford modules.

Lemma 2.16 *There exists an equivalence of categories*

$$\left\{ \begin{array}{c} \textit{Category of} \\ \textit{Cliff}(\mathbb{R}^{p,q})[G, \phi] - \textit{modules} \end{array} \right\} \longleftrightarrow \left\{ \begin{array}{c} \textit{Category of} \\ C_{p,q}[G, \phi] - \textit{modules} \end{array} \right\}. \tag{2.150}$$

In particular we have a natural isomorphism of groups

$$M^{p,q}(G, \phi) \cong \widehat{M}^{p,q}(G, \phi.) \tag{2.151}$$

The proof is just a generalization of the Atiyah's proof in [1] at the end of Sect. 4.

Proof Let M be a $C_{p,q}[G, \phi]$-module. We are going to define an action of $\mathrm{Cliff}(\mathbb{R}^{\cdot p,q})$ on M

$$\begin{cases} [e_j]m := e_j m & 1 \leq j \leq q \\ [e_j]m := i(e_j m) & q + 1 \leq j \leq q + p \end{cases} \tag{2.152}$$

where $[e_j]$ denotes a basis element in $\mathrm{Cliff}(\mathbb{R}^{p,q})$ and $i(e_j m)$ refers to the complex vector space structure of M. Then

$$[e_j]^2 m = e_j e_j m = e_j^2 m = -m \text{ for } 1 \leq j \leq q \tag{2.153}$$

$$[e_j]^2 m = i e_j i e_j m = -e_j^2 m = -m \text{ for } q + 1 \leq j \leq q + p. \tag{2.154}$$

So we have a well defined action of $\mathrm{Cliff}(\mathbb{R}^{p,q})$ on M. Now let us analyze the behavior of the action of (G, ϕ) with respect to this new action. For $g \in G, q + 1 \leq j \leq q + p$ and $m \in M$, we have

$$g([e_j]m) = g(i e_j m) \tag{2.155}$$

$$= (-1)^{\phi(g)} i g(e_j m) \tag{2.156}$$

$$= (-1)^{\phi(g)} i e_j g m \quad \text{(the action of } (G, \phi) \text{ commutes with } C_{p,q}) \tag{2.157}$$

$$= (-1)^{\phi(g)} [e_j] g m \tag{2.158}$$

$$= g[e_j] g m, \tag{2.159}$$

and for $g \in G$, $1 \leq j \leq q$ and $m \in M$, we have

$$g([e_j]m) = g(e_j m) \tag{2.160}$$

$$= e_j g m \tag{2.161}$$

$$= g[e_j] g m. \tag{2.162}$$

So we have a functor

$$\left\{ \begin{array}{c} \text{Category of} \\ \mathrm{Cliff}(\mathbb{R}^{p,q})[G, \phi] - \text{modules} \end{array} \right\} \longleftarrow \left\{ \begin{array}{c} \text{Category of} \\ C_{p,q}[G, \phi] - \text{modules} \end{array} \right\} \tag{2.163}$$

We can reverse the process. Let M be a $\mathrm{Cliff}(\mathbb{R}^{p,q})[G, \phi]$-module and define the action of $C_{p,q}$ by

$$\{e_j\}m = \begin{cases} e_j m & 1 \leq j \leq q \\ -i e_j m & q + 1 \leq j \leq q + p. \end{cases} \tag{2.164}$$

The negative sign makes these functors inverses of one another, in fact:

$$[\{e_j\}]m = \{e_j\}m = e_j m \text{ for } 1 \leq j \leq q \tag{2.165}$$

$$[\{e_j\}]m = i\{e_j\}m = -i^2 e_j m = e_j m \text{ for } q + 1 \leq j \leq q + p. \tag{2.166}$$

$$\square$$

The importance of this result is that we can pass from algebras with non-trivial involution to algebras with trivial involution. An application is the following theorem, which let us compute the K-theory of the spheres $S^{p,q}$.

Theorem 2.4 *The following diagram commutes*

$$M^{p,q+1}(G,\phi) \xrightarrow{\quad \text{res} \quad} M^{p,q}(G,\phi) \xrightarrow{\quad \alpha \quad} \mathbf{K}_G(D^{p,q},S^{p,q})$$

(2.167)

Proof The isomorphisms given in Lemma 2.16 are compatible with the restriction homomorphism because we have a commutative diagram of vector spaces

$$\begin{array}{ccc} \mathbb{R}^{q+1}\oplus i\mathbb{R}^p & \longleftarrow & \mathbb{R}^q\oplus i\mathbb{R}^p \\ \uparrow & & \uparrow \\ \mathbb{R}^{q+1}\oplus\mathbb{R}^p & \longleftarrow & \mathbb{R}^q\oplus\mathbb{R}^p \end{array}$$

(2.168)

where the morphisms $\mathbb{R}^q\oplus\mathbb{R}^p \longrightarrow \mathbb{R}^q\oplus i\mathbb{R}^p$ are given by

$$e_j \longmapsto \begin{cases} ie_j & 1\le j\le p \\ e_j & p+1\le j\le p+q. \end{cases}$$

(2.169)

$\square$

2.7 Coefficients of Magnetic Equivariant K-Theory

Now we restrict ourselves to the Clifford modules over C_q to calculate of the $\mathbf{K}_G$-theory of the spheres S^q without involution. We adopt the following notation from [21]:

$$M_{\mathbb{R}}^q(G,\phi) := M^{0,q}(G,\phi) = \widehat{M}^{0,q}(G,\phi)$$

(2.170)

One defines $M_{\mathbb{C}}^q(G,\phi)$ and $M_{\mathbb{H}}^q(G,\phi)$ similarly by replacing C_q by $C_q\otimes_{\mathbb{R}}\mathbb{C}$ and $C_q\otimes_{\mathbb{R}}\mathbb{H}$, respectively.

Denote by $M_{\mathbb{F}}^q$ the classical Clifford modules defined in [21] for $\mathbb{F}=\mathbb{R},\mathbb{C},\mathbb{H}$.

Proposition 2.4 *There exists an isomorphism*

$$M_{\mathbb{R}}^q(G,\phi) \cong \big(\mathbf{R}(G,\mathbb{R})\otimes M_{\mathbb{R}}^q\big)\oplus\big(\mathbf{R}(G,\mathbb{C})\otimes M_{\mathbb{C}}^q\big)\oplus\big(\mathbf{R}(G,\mathbb{H})\otimes M_{\mathbb{H}}^q\big).$$

(2.171)

Proof Let $M = M_0\oplus M_1\in M_{\mathbb{R}}^q(G,\phi)$. Consider the following decomposition, similar to that given in Proposition 2.1

$$M_0 \oplus M_1 \cong \bigoplus_{V \in \mathbf{Irrep}(G)} \left(V \otimes_{\mathrm{End}_{\mathbf{Rep}(G)}(V)} \mathrm{Hom}_{\mathbf{Rep}(G)}(V, M_0 \oplus M_1) \right) \tag{2.172}$$

By the Schur lemma (Lemma 1.1), the set $\mathrm{Hom}_{\mathbf{R}(G)}(V, M_0 \oplus M_1)$ is a vector space over $\mathbb{R}$, $\mathbb{C}$ or $\mathbb{H}$, depending on the type of the representation V with trivial action of (G, ϕ). Therefore $\mathrm{Hom}_{\mathbf{R}(G)}(V, M_0 \oplus M_1) \in M^q_{\mathrm{End}_{\mathbf{R}(G)}}$. $\square$

Proposition 2.5 *There following sequence is exact*

$$M^{q+1}_{\mathbb{R}}(G, \phi) \xrightarrow{\ \mathrm{res}\ } M^q_{\mathbb{R}}(G, \phi) \xrightarrow{\ \alpha\ } \mathbf{K}_G(D^q, S^{q-1}) \longrightarrow 0. \tag{2.173}$$

Proof Using Propositions 2.1 and 2.4 we see the sequence splits in three sequences

$$M^{q+1}_{\mathbb{F}} \longrightarrow M^q_{\mathbb{F}} \xrightarrow{\ \alpha\ } KF(D^q, S^{q-1}) \longrightarrow 0 \tag{2.174}$$

for $(\mathbb{F}, KF) = (\mathbb{R}, KO), (\mathbb{C}, KU), (\mathbb{H}, KSp)$. Atiyah and Segal proved the exactness of these three sequences in [2, Proposition 8.2]. $\square$

So we have a way to compute the K-theory of the point in terms of the Clifford modules

$$\mathbf{K}_G(D^q, S^{q-1}) \cong M^q_{\mathbb{R}}(G, \phi)/M^{q+1}_{\mathbb{R}}(G, \phi).$$

We now compute the coefficients of the $\mathbf{K}^*_G$-theory, where the representations of the magnetic groups will appear.

Theorem 2.5 (Coefficients of $\mathbf{K}^*_G$) *Let (G, ϕ) be a finite magnetic group. Denote by $n_{\mathbb{F}}$ the number of isomorphism classes of irreducible representations of (G, ϕ) of real, complex or quaternionic type for $\mathbb{F} = \mathbb{R}, \mathbb{C}, \mathbb{H}$. Then the coefficients of the $\mathbf{K}^*_G$-theory are given by*

$$\begin{array}{c|c}
q & \mathbf{K}^{-q}_G(pt) \\
\hline
0 & \mathbb{Z}^{n_{\mathbb{R}}} \oplus \mathbb{Z}^{n_{\mathbb{C}}} \oplus \mathbb{Z}^{n_{\mathbb{H}}} \\
1 & (\mathbb{Z}/2)^{n_{\mathbb{R}}} \\
2 & (\mathbb{Z}/2)^{n_{\mathbb{R}}} \oplus \mathbb{Z}^{n_{\mathbb{C}}} \\
3 & 0 \\
4 & \mathbb{Z}^{n_{\mathbb{R}}} \oplus \mathbb{Z}^{n_{\mathbb{C}}} \oplus \mathbb{Z}^{n_{\mathbb{H}}} \\
5 & (\mathbb{Z}/2)^{n_{\mathbb{H}}} \\
6 & \mathbb{Z}^{n_{\mathbb{C}}} \oplus (\mathbb{Z}/2)^{n_{\mathbb{H}}} \\
7 & 0
\end{array} \tag{2.175}$$

and these groups are of period 8 in q. Alternatively, we have the canonical isomorphism:

$$\mathbf{K}^*_G(pt) \cong KO^*(pt)^{\oplus n_{\mathbb{R}}} \oplus KU^*(pt)^{\oplus n_{\mathbb{C}}} \oplus KSp^*(pt)^{\oplus n_{\mathbb{H}}} \tag{2.176}$$

Proof Using the decomposition of $M_{\mathbb{R}}^q(G, \phi)$ from Proposition 2.4, we have the following commutative diagram:

$$M^{q+1}(G,\phi) \xrightarrow{\cong} \bigoplus_{\mathbb{F}\in\{\mathbb{R},\mathbb{C},\mathbb{H}\}} \left(\mathbf{R}(G,\mathbb{F}) \otimes M_{\mathbb{F}}^{q+1}\right) \xrightarrow{\cong} \bigoplus_{\mathbb{F}\in\{\mathbb{R},\mathbb{C},\mathbb{H}\}} \left(\mathbb{Z}^{n_{\mathbb{F}}} \otimes M_{\mathbb{F}}^{q+1}\right) \qquad (2.177)$$

$$\downarrow^{\mathrm{res}} \qquad\qquad \downarrow^{\oplus\mathrm{res}} \qquad\qquad \oplus\mathrm{res}\downarrow$$

$$M^q(G,\phi) \xrightarrow{\cong} \bigoplus_{\mathbb{F}\in\{\mathbb{R},\mathbb{C},\mathbb{H}\}} \left(\mathbf{R}(G,\mathbb{F}) \otimes M_{\mathbb{F}}^q\right) \xrightarrow{\cong} \bigoplus_{\mathbb{F}\in\{\mathbb{R},\mathbb{C},\mathbb{H}\}} \left(\mathbb{Z}^{n_{\mathbb{F}}} \otimes M_{\mathbb{F}}^q\right)$$

so we can conclude that

$$M^q(G,\phi)/\mathrm{res}(M^{q+1}(G,\phi)) \cong \bigoplus_{\mathbb{F}\in\{\mathbb{R},\mathbb{C},\mathbb{H}\}} \left(\mathbb{Z}^{n_{\mathbb{F}}} \otimes M_{\mathbb{F}}^q/\mathrm{res}(M_{\mathbb{F}}^{q+1})\right) = \bigoplus_{\mathbb{F}\in\{\mathbb{R},\mathbb{C},\mathbb{H}\}} \left(M_{\mathbb{F}}^q/\mathrm{res}(M_{\mathbb{F}}^{q+1})\right)^{n_{\mathbb{F}}}.$$
$$(2.178)$$

The groups $M_{\mathbb{F}}^q/\mathrm{res}(M_{\mathbb{F}}^{q+1})$ give the complex, real and quaternionic K-theory and have been computed in [1, 21]. The decomposition follows. $\qquad\square$

2.8 Twistings, Degree Shift and Periodicity

In quantum mechanics, symmetry groups act projectively; therefore it is imperative that we understand how these type of actions relate to the magnetic equivariant K-theory groups. We will focus only on projective actions that come from sign representations of copies of the group $\mathbb{Z}/2$.

The setup is the following. Start with a magnetic group (G, ϕ) and consider a magnetic central extension $(\widetilde{G}, \widetilde{\phi})$ of (G, ϕ) by an abelian group $A \cong (\mathbb{Z}/2)^k$.

If $\widetilde{G}_0$ denotes the core of $\widetilde{G}$ we have the following diagram of group extensions:

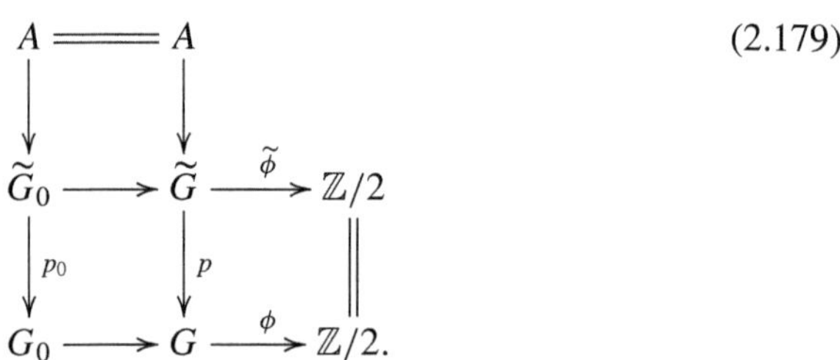

$$(2.179)$$

Such extensions are classified by the cohomology group $H^2(G, A)$, but instead of working with a cocycle representing the extension, we will work with the group extension itself.

Definition 2.11 Let X be a G-space, $(\widetilde{G}, \widetilde{\phi})$ a magnetic group extension of (G, ϕ) by the abelian group $A \cong (\mathbb{Z}/2)^k$ as above, and $\chi \in \widehat{A} := \mathrm{Hom}(A, \mathrm{U}(1))$ an irreducible representation of A.

Let $^{(\widetilde{G},\chi)}\mathbf{K}^*_G(X)$, the $(\widetilde{G},\chi)$**-twisted magnetic G-equivariant K-theory groups of** X, be the direct summand of the magnetic $\widetilde{G}$-equivariant K-theory of X

$$^{(\widetilde{G},\chi)}\mathbf{K}^*_G(X) \subset \mathbf{K}^*_{\widetilde{G}}(X) \tag{2.180}$$

generated by magnetic $(\widetilde{G},\widetilde{\phi})$-equivariant vector bundles on which the action of A on the fibers matches the one of the character χ (cf. [22, 23]).

Remark 2.8 Since X is a G-space, the induced action of $\widetilde{G}$ on X has the property that A acts trivially. The group A being abelian allows us to decompose any magnetic $(\widetilde{G},\widetilde{\phi})$-equivariant vector bundle by the action of A on the fibers

$$\mathbf{K}_{\widetilde{G}}(X) \cong \bigoplus_{\chi \in \widehat{A}} {}^{(\widetilde{G},\chi)}\mathbf{K}^*_G(X). \tag{2.181}$$

According to the Definition 2.11 we define

$$^{(\widetilde{G},\chi)}\mathbf{Rep}(G),\quad {}^{(\widetilde{G},\chi)}\mathbf{R}(G),\quad {}^{(\widetilde{G},\chi)}\mathbf{R}(G,\mathbb{F}),\quad {}^{(\widetilde{G},\chi)}\mathbf{Irrep}(G,\mathbb{F}),\quad {}^{(\widetilde{G},\chi)}\mathbf{Vect}_G(X) \tag{2.182}$$

as the category of $(\widetilde{G},\widetilde{\phi})$ representations where A acts by the character $\chi \in \widehat{A}$, its Grothendieck group, the ones of type $\mathbb{F}$, the irreducible ones, the irreducible ones of type $\mathbb{F}$ and the magnetic $(\widetilde{G},\widetilde{\phi})$-equivariant vector bundles over X where A acts by the character χ, respectively.

Example 2.8 Of particular relevance is the case on which $(G,\phi) = (\mathbb{Z}/2, id)$ and $(\widetilde{G},\widetilde{\phi}) = (\mathbb{Z}/4, \mathrm{mod}\,2)$. Here $A \cong \mathbb{Z}/2$ and we have two characters. If X is a $\mathbb{Z}/2$-space, any magnetic $(\mathbb{Z}/4, \mathrm{mod}\,2)$-equivariant vector bundle over X decomposes as a direct sum of two bundles, one on which A acts trivially, and therefore could be thought as a magnetic $(\mathbb{Z}/2, id)$-equivariant vector bundle, and one on which $A \cong \mathbb{Z}/2$ acts by multiplication by -1. The first type is known as Atiyah's Real vector bundles. The second type are equivalent to what is known in the literature as quaternionic vector bundles (Dupont [12] calls them symplectic bundles, but we will leave the name symplectic for the case where the space X has no action of $\mathbb{Z}/2$). Hence:

$$\mathbf{K}^*_{\mathbb{Z}/4}(X) \cong K\mathbb{R}^*(X) \oplus K\mathbb{H}^*(X), \tag{2.183}$$

see Example 2.3. If we denote by σ the sign representation of $A = \mathbb{Z}/2$, then the $(\mathbb{Z}/4, \sigma)$-twisted magnetic $\mathbb{Z}/2$-equivariant K-theory over $\mathbb{Z}/2$-spaces is clearly equivalent to quaternionic K-theory:

$$^{(\mathbb{Z}/4,\sigma)}\mathbf{K}^*_{\mathbb{Z}/2}(X) \cong K\mathbb{H}^*(X). \tag{2.184}$$

The degree shift in magnetic equivariant K-theory comes from the $\mathbb{Z}/2$-central extension of any magnetic group (G, ϕ) defined by the pullback of $\mathbb{Z}/4$ under ϕ. This particular choice of magnetic group is

$$(\widehat{G}, \widehat{\phi}) := (G \times_{\mathbb{Z}/2} \mathbb{Z}/4, mod_2 \circ \pi_2) \tag{2.185}$$

where we have that:

$$\widehat{G} = \phi^*(\mathbb{Z}/4) = G \times_{\mathbb{Z}/2} \mathbb{Z}/4 = \{(g, m) \in G \times \mathbb{Z}/4 : \phi(g) = mod_2(m)\}, \tag{2.186}$$

$$\text{and} \quad \widehat{\phi}(g, m) = mod_2 \circ \pi_2(g, m) = mod_2(m), \tag{2.187}$$

and whose core is $\widehat{G}_0 \cong G_0 \times \mathbb{Z}/2$. Diagrammatically we have:

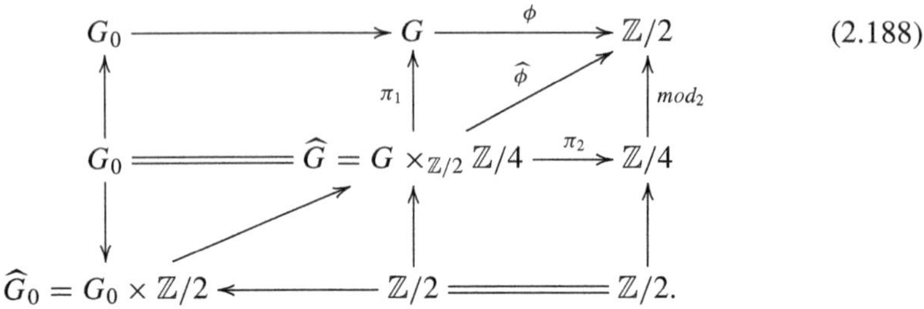

$$\tag{2.188}$$

Denote by $\widehat{\sigma} := \phi^*\sigma$ the sign representation on $\mathbb{Z}/2 = \mathrm{Ker}(\widehat{G} \to G)$ coming from σ on $\mathbb{Z}/2 = \mathrm{Ker}(\mathbb{Z}/4 \to \mathbb{Z}/2)$. We claim the following degree shift isomorphism.

Theorem 2.6 (Degree shift isomorphism) *For X a G-space and (G, ϕ) a magnetic group, then there is a natural isomorphism*

$$\mathbf{K}_G^*(X) \xrightarrow{\cong} {}^{(\widehat{G}, \widehat{\sigma})}\mathbf{K}_G^{*+4}(X). \tag{2.189}$$

Proof We begin by denoting the twisted equivariant magnetic K-theory by ${}^{\widehat{G}}\mathbf{K}_G^*(X)$, since there is only one non-trivial representation of $\mathbb{Z}/2 = \mathrm{Ker}(\widehat{G} \to G)$.

We will generalize the proof of the degree shift isomorphism that was shown by Dupont [12] to the equivariant setup. We will follow closely the steps of his original proof.

Take $(B^{3,0}, S^{3,0})$ the pair consisting of the unit ball and the unit sphere in $\mathbb{R}^{3,0}$. Take $\mathbb{H}$ the quaternions and denote by $\mathbb{P}(\mathbb{H})$ its complex projectivization together with the canonical complex bundle $H_{sp} \to \mathbb{P}(\mathbb{H})$ over it. Note that the magnetic group $\mathbb{Z}/4$ acts on $\mathbb{H} \cong \mathbb{C} \oplus \mathbb{C}$ via the matrix $\left(\begin{smallmatrix} 0 & 1 \\ -1 & 0 \end{smallmatrix}\right)\mathbb{K}$ and therefore the complex bundle $H_{sp} \to \mathbb{P}(\mathbb{H})$ becomes magnetic $\mathbb{Z}/4$-equivariant over $\mathbb{P}(\mathbb{H})$. The induced action of $\mathbb{Z}/4$ on $\mathbb{P}(\mathbb{H})$ is trivial on $\mathrm{Ker}(\mathbb{Z}/4 \to \mathbb{Z}/2)$ and moreover this induced $\mathbb{Z}/2$ action on $\mathbb{P}(\mathbb{H})$ is free.

Therefore $\mathbb{P}(\mathbb{H})$ as a $\mathbb{Z}/2$ space is equivalent to $S^{3,0}$ and H_{sp} is a $\mathbb{Z}/4$-twisted magnetic $\mathbb{Z}/2$-equivariant line bundle over $S^{3,0}$:

$$[H_{sp}] \in {}^{\mathbb{Z}/4}\mathbf{K}^0_{\mathbb{Z}/2}(S^{3,0}) = K\mathbb{H}^0(S^{3,0}). \tag{2.190}$$

The long exact sequence associated to the pair $(X \times B^{3,0}, X \times S^{3,0})$ yields, by [1, Corollary 3.8] the short exact sequence

$$0 \longrightarrow \mathbf{K}^*_G(X) \xrightarrow{\pi^*} \mathbf{K}^*_G(X \times S^{3,0}) \xrightarrow{\delta} \mathbf{K}^{*+4}_G(X) \to 0 \tag{2.191}$$

where G is acting on $B^{3,0}$ via ϕ, on the product $X \times B^{3,0}$ diagonally, and $\pi : X \times S^{3,0} \to X$ is the projection.

This short exact sequence applied to ${}^{\mathbb{Z}/4}\mathbf{K}_{\mathbb{Z}/2}$ gives the sequence [12, Eq. 5]

$$0 \longrightarrow {}^{\mathbb{Z}/4}\mathbf{K}^0_{\mathbb{Z}/2}(*) \xrightarrow{\pi^*} {}^{\mathbb{Z}/4}\mathbf{K}^0_{\mathbb{Z}/2}(S^{3,0}) \xrightarrow{\delta} {}^{\mathbb{Z}/4}\mathbf{K}^4_{\mathbb{Z}/2}(*) \to 0 \tag{2.192}$$

where $d_0 := \delta H_{sp}$ is the generator of ${}^{\mathbb{Z}/4}\mathbf{K}^4_{\mathbb{Z}/2}(*) = KSp^4(*) = \mathbb{Z}$, and $d_0^2 \in \mathbf{K}^8_{\mathbb{Z}/2}(*) = K\mathbb{R}^8(*)$ is also the generator [12, Eq. 6].

Consider now the map

$$\mathbf{K}_G(X) \times {}^{\mathbb{Z}/4}\mathbf{K}_{\mathbb{Z}/2}(B^{4,0}, S^{4,0}) \to {}^{\widehat{G}}\mathbf{K}_G(X \times B^{4,0}, X \times S^{4,0})$$
$$(E, \delta H_{sp}) \mapsto \pi_1^* E \otimes \pi_2^* \delta H_{sp} \tag{2.193}$$

where π_1 and π_2 are the projections and δH_{sp} denotes the generator of ${}^{\mathbb{Z}/4}\mathbf{K}^4_{\mathbb{Z}/2}(*)$ defined above. Note that the bundle $\pi_1^* E \otimes \pi_2^* \delta H_{sp}$ is $\widehat{G}$-equivariant since both E and δH_{sp} are also $\widehat{G}$-equivariant, the former through the induced action of G and the latter through the induced action of $\mathbb{Z}/4$. Note moreover that $\mathbb{Z}/2 = \ker(\widehat{G} \to G)$ acts by multiplication by -1 on $\pi_1^* E \otimes \pi_2^* \delta H_{sp}$ and therefore it belongs to the $\widehat{G}$-twisted magnetic G-equivariant K-theory group.

We summarize the construction of Eq. (2.193) as the desired degree shifting map

$$\mathbf{K}^0_G(X) \to {}^{\widehat{G}}\mathbf{K}^4_G(X)$$
$$E \mapsto E \cdot d_0. \tag{2.194}$$

We claim that this homomorphism is an isomorphism because when applied twice it induces the 8-periodicity isomorphism:

$$\mathbf{K}^0_G(X) \xrightarrow{\cong} \mathbf{K}^8_G(X)$$
$$E \mapsto E \cdot d_0^2, \tag{2.195}$$

where d_0^2 is the generator of $\mathbf{K}^8_{\mathbb{Z}/2}(*)$, as it is shown in [12], thus inducing the 8-periodicity of Theorem 2.7. The fact that the homomorphism of Eq. (2.195) is an isomorphism follows from the Mayer-Vietoris sequence and the five lemma.   □

Following the notation of Theorem 2.6, applying the degree isomorphism twice delivers the isomorphism

$$^{(\tilde{G},\hat{\sigma})}\mathbf{K}^{*+4}_G(X) \xrightarrow{\cong} {}^{(\widehat{\widehat{G}},\widehat{\hat{\sigma}}+\hat{\sigma})}\mathbf{K}^{*+8}_G(X), \tag{2.196}$$

where $(\widehat{\widehat{G}}, \widehat{\widehat{\phi}})$ is the magnetic group

$$\widehat{\widehat{G}} = \{(g,m,n) \in G \times \mathbb{Z}/4 \times \mathbb{Z}/4 \colon \phi(g) = mod_2(m) = mod_2(n)\} \tag{2.197}$$

with $\widehat{\widehat{\phi}}(g,m,n) = mod_2(n)$, and $\widehat{\hat{\sigma}} + \hat{\sigma}$ the irreducible representation of

$$(\mathbb{Z}/2)^2 \cong \langle (1_G,2,0),(1_G,0,2)\rangle = \mathrm{Ker}(\widehat{\widehat{G}} \to G) \tag{2.198}$$

that maps both $(1_G,2,0)$ and $(1_G,0,2)$ to -1. Then we have the following simple result.

Lemma 2.17 *There is a natural isomorphism*

$$\mathbf{K}^*_G(X) \xrightarrow{\cong} {}^{(\widehat{\widehat{G}},\widehat{\hat{\sigma}}+\hat{\sigma})}\mathbf{K}^*_G(X). \tag{2.199}$$

Proof Take a magnetic G-equivariant vector bundle E over X and endow this same vector bundle E but with the following action of $\widehat{\widehat{G}}$. For $e \in E$ and $(g,m,n) \in \widehat{\widehat{G}}$ let

$$(g,m,n) \cdot e := (-1)^{\frac{m+n}{2}} g \cdot e. \tag{2.200}$$

The action is well defined because $m+n$ is even. Furthermore, both $(1_G,2,0)$ and $(1_G,0,2)$ act by multiplication by -1. With this action of $\widehat{\widehat{G}}$ the vector bundle E is a $(\widehat{\widehat{G}},\widehat{\hat{\sigma}}+\hat{\sigma})$-twisted magnetic G-equivariant vector bundle and therefore we get the following isomorphism of categories

$$\mathbf{Vec}_G(X) \xrightarrow{\cong} {}^{(\widehat{\widehat{G}},\widehat{\hat{\sigma}}+\hat{\sigma})}\mathbf{Vec}_G(X). \tag{2.201}$$

The isomorphism in K-theories follows.   □

Therefore the degree shift homomorphism gives us isomorphisms

$$\mathbf{K}^*_G(X) \xrightarrow{\cong} {}^{(\tilde{G},\hat{\sigma})}\mathbf{K}^{*+4}_G(X) \xrightarrow{\cong} \mathbf{K}^{*+8}_G(X) \tag{2.202}$$

whose composition $\mathbf{K}_G^*(X) \xrightarrow{\cong} \mathbf{K}_G^{*+8}(X)$ is multiplying by d_0^2, or alternatively, the canonical isomorphism given by the Morita equivalence of the Clifford algebras $C_{p,q}$ and $C_{p+8,q}$.

An interesting consequence of the degree shift isomorphism of Theorem 2.6 is the following.

Proposition 2.6 *If the short exact sequence $\mathbb{Z}/2 \to \widehat{G} \to G$ splits (namely $\widehat{G} \cong G \times \mathbb{Z}/2$ is a trivial extension of G), then the magnetic equivariant K-theory $\mathbf{K}_G^*(X)$ of the G-space X is 4-periodic.*

Proof Let $\alpha : G \to \widehat{G}$ be the splitting and $G \times \mathbb{Z}/2 \to \widehat{G}$, $(g, m) \mapsto \alpha(g) \circ (1_G, m)$ the induced isomorphism. Then any $(\widehat{G}, \widehat{\sigma})$-twisted magnetic G-equivariant vector bundle translates to a magnetic $(G \times \mathbb{Z}/2, sgn)$-twisted G-equivariant vector bundle where the $\mathbb{Z}/2$ component acts on the fibers by the sign representation (multiplication by -1). Forgetting this action of $\mathbb{Z}/2$ gives the desired isomorphism

$$^{(\widehat{G},\widehat{\sigma})}\mathbf{K}_G^*(X) = {}^{(G \times \mathbb{Z}/2, sgn)}\mathbf{K}_G^*(X) \cong \mathbf{K}_G^*(X). \tag{2.203}$$

The degree shift isomorphism of Theorem 2.6 provides the desired periodicity of degree 4:

$$\mathbf{K}_G^*(X) \xrightarrow{\cong} {}^{(\widehat{G},\widehat{\sigma})}\mathbf{K}_G^{*+4}(X) \xrightarrow{\cong} \mathbf{K}_G^{*+4}(X). \tag{2.204}$$

$\square$

Example 2.9 Take the cyclic magnetic group $\mathbb{Z}/2n$ for $n > 1$. Then the extension

$$\widehat{\mathbb{Z}/2n} = \{(a, b) \in \mathbb{Z}/2n \times \mathbb{Z}/4 : a + b = 0 \ \mathrm{mod}\, 2\} \tag{2.205}$$

splits with isomorphism

$$\widehat{\mathbb{Z}/2n} \cong \mathbb{Z}/2n \times \mathbb{Z}/2, \quad (a, b) \mapsto \left(a, \mathrm{mod}\, 2\left(\frac{a+b}{2}\right)\right). \tag{2.206}$$

Therefore the magnetic $\mathbb{Z}/2n$-equivariant K-theory $\mathbf{K}_{\mathbb{Z}/2n}^*(X)$ is 4-periodic.

Now we are ready to prove the 8-periodicity of the Magnetic Equivariant K-theory. This will be based on the construction of the degree shift homomorphism but defined at the level of Grothendieck group of isomorphism classes of magnetic $C_{p,q}[G, \phi]$-modules $\widehat{M}^{p,q}(G, \phi)$ of Definition 2.10. The results presented in what follows have been constructed from a combination of results appearing in [21, Sects. 4 and 5], [1, Sect. 4], [2, Sect. 8], [6, 24, I Sect. 3] but written in a convenient base which further clarifies their relevance. We need first some definitions.

Take $(\widetilde{G}, \widetilde{\phi})$ a magnetic group extension of (G, ϕ) by the abelian group $A \cong (\mathbb{Z}/2)^k$ and $\chi \in \hat{A} := \mathrm{Hom}(A, U(1))$ an irreducible representation of A as in Definition 2.11.

Definition 2.12 Let ${}^{(\widetilde{G},\chi)}\widehat{M}^{p,q}(G,\phi)$ be the Grothendieck group of isomorphism classes of $C_{p,q}[\widehat{G},\widehat{\phi}]$-modules as in Definition 2.10, where A acts by the character χ.

Clearly we have the inclusion ${}^{(\widetilde{G},\chi)}\widehat{M}^{p,q}(G,\phi) \subset \widehat{M}^{p,q}(\widetilde{G},\widetilde{\phi})$.

For the canonical $\mathbb{Z}/2$ extension $\widehat{G}$ of G defined in Eq. (2.186) by the pullback of $\phi : G \to \mathbb{Z}/2$, denote

$$
{}^{\widehat{G}}\widehat{M}^{p,q}(G,\phi) := {}^{(\widehat{G},\widehat{\sigma})}\widehat{M}^{p,q}(G,\phi) \tag{2.207}
$$

where σ is the non-trivial character of $\mathbb{Z}/2$ as defined in Theorem 2.6. Exactly as in Lemma 2.17 we get a canonical isomorphism whenever the twist is done twice

$$
\widehat{M}^{p,q}(G,\phi) \xrightarrow{\cong} {}^{(\widehat{\widehat{G}},\widehat{\widehat{\sigma}}+\widehat{\sigma})}\widehat{M}^{p,q}(G,\phi). \tag{2.208}
$$

We need to recall the following identities for the Clifford algebras $C_{p,q}$ defined in Eq. (2.128) whose proof can be found in [21, Sect. 4] and [24, I Sect. 3]:

$$
C_{p+r,q+s} \cong C_{p,q}\hat{\otimes}_{\mathbb{R}}C_{r,s} \qquad\qquad \mathbb{H} \cong C_{0,2} \tag{2.209}
$$

$$
\mathbb{H}(2) \cong C_{0,4} \cong C_{4,0} \qquad\qquad \mathbb{R}(16) \cong C_{0,8} \tag{2.210}
$$

where $\mathbb{F}(n)$ denotes the algebra of $n \times n$ matrices over $\mathbb{F}$ and $\hat{\otimes}$ denotes the graded tensor product.

Following [6], we define the graded $C_{0,2}$-module $\Delta := \mathbb{C}_0 \oplus \mathbb{C}_1$ with $\mathbb{C}_0$ denoting the complex numbers in degree 0 and $\mathbb{C}_1$ the complex numbers in degree 1. The action of the Clifford algebra $C_{0,2}$ in terms of the generators e_1 and e_2 is given by the following matrices:

$$
e_1 = \begin{pmatrix} 0 & 1 \\ -1 & 0 \end{pmatrix}, \quad e_2 = \begin{pmatrix} 0 & i \\ i & 0 \end{pmatrix}. \tag{2.211}
$$

The module Δ can moreover be endowed with a degree 1 automorphism $c = \begin{pmatrix} 0 & 1 \\ 1 & 0 \end{pmatrix}\mathbb{K}$ which is antilinear with respect of the $\mathbb{C}$-module structure of Δ and that it commutes (in the graded sense) with e_1 and e_2.

The graded tensor product $\Delta^{\hat{\otimes}2} := \Delta\hat{\otimes}\Delta$ becomes a graded module of $C_{0,4} \cong C_{0,2}\hat{\otimes}C_{0,2}$ (see [21, Sect. 6] for the definition of the graded tensor product and [24, I §3] for the graded tensor product of Clifford algebras) and the operator $c\hat{\otimes}c$ becomes a degree 0 operator, which is antilinear in $\mathbb{C}$ that squares to -1. Following the commuting rules of graded compositions and graded tensor products we have

$$
(c\hat{\otimes}c) \circ (c\hat{\otimes}c) = -(c \circ c)\hat{\otimes}(c \circ c) = -\mathrm{Id}. \tag{2.212}
$$

The operator $c\hat{\otimes}c$ defines a magnetic $\mathbb{Z}/4$ action on $\Delta^{\hat{\otimes}2}$ which commutes with the $C_{0,4}$-actions thus defining an element in $^{\mathbb{Z}/4}\widehat{M}^{0,4}(\mathbb{Z}/2, id)$ as in Eq. (2.207), where $id : \mathbb{Z}/2 \to \mathbb{Z}/2$ is the identity isomorphism and $\mathbb{Z}/4 = \widehat{\mathbb{Z}/2}$.

Proposition 2.7 *The homomorphism*

$$\beta : \widehat{M}^{p,q}(G, \phi) \overset{\cong}{\to} {}^{\widehat{G}}\widehat{M}^{p,q+4}(G, \phi)$$

$$M \mapsto M\hat{\otimes}\Delta^{\hat{\otimes}2} \tag{2.213}$$

induces an isomorphism. Applying it twice induces the 8-periodicity

$$(\beta \circ \beta) : \widehat{M}^{p,q}(G, \phi) \overset{\cong}{\to} \widehat{M}^{p,q+8}(G, \phi). \tag{2.214}$$

Proof The tensor product $M\hat{\otimes}\Delta^{\hat{\otimes}2}$ becomes a $\widehat{G}$-representation since M has an induced $\widehat{G}$-action by the quotient homomorphism $\widehat{G} \to G$ and $\Delta^{\hat{\otimes}2}$ has an induced action via the homomorphism $\widehat{G} \to \mathbb{Z}/4$. This is a $(\widehat{G}, \widehat{\sigma})$-twisted magnetic representation of G which moreover commutes with the action of $C_{p,q+4} \cong C_{p,q}\hat{\otimes}C_{0,4}$.

Tensoring twice with $\Delta^{\hat{\otimes}2}$ is equivalent to tensoring with $\Delta^{\hat{\otimes}4}$. Now, this is a $C_{0,8}$-module whose $\mathbb{C}$-antilinear map $c^{\hat{\otimes}4}$ is of degree 0 and squares to 1. The complex dimension of $\Delta^{\hat{\otimes}4}$ is 16 and the $c^{\hat{\otimes}4}$-invariant part (its real part) has real dimension 16. Since the smallest irreducible module of $C_{0,8} \cong \mathbb{R}(16)$ is of dimension 16, we can conclude that $\Delta^{\hat{\otimes}4}$ is irreducible.

The composition of β twice defines a homomorphism

$$\beta : \widehat{M}^{p,q}(G, \phi) \overset{\cong}{\to} {}^{(\widehat{\widehat{G}}, \widehat{\widehat{\sigma}}+\widehat{\sigma})}\widehat{M}^{p,q+8}(G, \phi), \tag{2.215}$$

but by Lemma (2.17) we know that $^{(\widehat{\widehat{G}}, \widehat{\widehat{\sigma}}+\widehat{\sigma})}\widehat{M}^{p,q+8}(G, \phi) \cong \widehat{M}^{p,q+8}(G, \phi)$. Therefore the assignment $(\beta \circ \beta)(M) = M\hat{\otimes}\Delta^{\hat{\otimes}4}$ induces the desired isomorphism groups $(\beta \circ \beta) : \widehat{M}^{p,q}(G, \phi) \overset{\cong}{\to} \widehat{M}^{p,q+8}(G, \phi)$ which realizes the well-known 8-periodicity of Clifford module theory [24, Theorem 4.3]. $\qquad\square$

Choosing a base for the tensor product in $\Delta^{\hat{\otimes}2}$ we can give a explicit description of this $\mathbb{Z}/4$-twisted $C_{0,4}[\mathbb{Z}/2, id]$-module. As a graded vector space it is $\mathbb{C}_0 \oplus \mathbb{C}_1 \oplus \mathbb{C}_1 \oplus \mathbb{C}_0$, the action of the generators of the Clifford algebra are the following matrices:

$$e_1 = \begin{pmatrix} 0 & 0 & 1 & 0 \\ 0 & 0 & 0 & 1 \\ -1 & 0 & 0 & 0 \\ 0 & -1 & 0 & 0 \end{pmatrix} \quad e_2 = \begin{pmatrix} 0 & 0 & i & 0 \\ 0 & 0 & 0 & i \\ i & 0 & 0 & 0 \\ 0 & i & 0 & 0 \end{pmatrix} \quad e_3 = \begin{pmatrix} 0 & 1 & 0 & 0 \\ -1 & 0 & 0 & 0 \\ 0 & 0 & 0 & -1 \\ 0 & 0 & 1 & 0 \end{pmatrix} \quad e_4 = \begin{pmatrix} 0 & i & 0 & 0 \\ i & 0 & 0 & 0 \\ 0 & 0 & 0 & -i \\ 0 & 0 & -i & 0 \end{pmatrix}$$

$$\tag{2.216}$$

and the $\mathbb{Z}/4$-magnetic action $c^{\hat{\otimes}2}$ is given by the following matrix:

$$c^{\hat{\otimes}2} = \begin{pmatrix} 0 & 0 & 0 & 1 \\ 0 & 0 & 1 & 0 \\ 0 & -1 & 0 & 0 \\ -1 & 0 & 0 & 0 \end{pmatrix} \mathbb{K}. \tag{2.217}$$

We claim that the homomorphism β of Proposition 2.7 is compatible with the Atiyah-Bott-Shapiro homomorphism defined in Eq. (2.142) and Theorem (2.4). For this, we first need to define an alternative degree shift homomorphism using Clifford modules. Here we are going to use the alternative definition of the K-theory groups using complexes of vector bundles presented in Definition 2.14.

Consider (X, A) a G-pair of locally compact spaces. An element in $\mathbf{K}_G(X, A)$ is a triple (E^0, E^1, φ) where E^0 and E^1 are magnetic (G, ϕ)-vector bundles, $\varphi : E^0 \to E^1$ is a homomorphism of (G, ϕ)-vector bundles and its restriction $\varphi|_A : E^0|_A \xrightarrow{\cong} E^1|_A$ to A is an isomorphism. By Lemma 2.6 there exists an extension of $\overline{\varphi} : E^1 \to E^0$ of $(\varphi|_A)^{-1} : E^1|_A \xrightarrow{\cong} E^0|_A$ and we can consider the homomorphism

$$\varphi \oplus \overline{\varphi} : E^0 \oplus E^1 \to E^1 \oplus E^0 \tag{2.218}$$

as a degree 1 map from the graded vector bundle $E^* = (E^0, E^1)$.

Definition 2.13 The degree shift homomorphism (using Clifford modules) is the assignment

$$\gamma : \mathbf{K}_G(X, A) \to {}^{\widehat{G}}\mathbf{K}_G \left(B^{0,4} \times X, (S^{0,4} \times X) \cup (B^{0,4} \times A) \right) = {}^{\widehat{G}}\mathbf{K}_G^{0,4}(X, A)$$

$$(E^0, E^1, \varphi) \mapsto \left(B^{0,4} \times (E^* \hat{\otimes} \Delta^{\hat{\otimes}2})_0, \, B^{0,4} \times (E^* \hat{\otimes} \Delta^{\hat{\otimes}2})_1, \, \psi \right) \tag{2.219}$$

where $(E^* \hat{\otimes} \Delta^{\hat{\otimes}2})_i$ is the degree i part of the graded tensor product $E^* \hat{\otimes} \Delta^{\hat{\otimes}2}$ and

$$\psi : B^{0,4} \times (E^* \hat{\otimes} \Delta^{\hat{\otimes}2})_0 \to B^{0,4} \times (E^* \hat{\otimes} \Delta^{\hat{\otimes}2})_1 \tag{2.220}$$

is the homomorphism that on $x \in B^{0,4}$ is defined as

$$\psi_x(z_0 \otimes \xi_0) = \varphi(z_0) \otimes \xi_0 + z_0 \otimes x\xi_0 \tag{2.221}$$
$$\psi_x(z_1 \otimes \xi_1) = -\overline{\varphi}(z_1) \otimes \xi_1 - z_1 \otimes x\xi_1 \tag{2.222}$$

with $z_0 \otimes \xi_0 \in E^0 \otimes (\Delta^{\hat{\otimes}2})_0$ and $z_1 \otimes \xi_1 \in E^1 \otimes (\Delta^{\hat{\otimes}2})_1$, where the decomposition by degree is

$$(E^* \hat{\otimes} \Delta^{\hat{\otimes}2})_0 \cong \left(E^0 \otimes (\Delta^{\hat{\otimes}2})_0 \right) \oplus \left(E^1 \otimes (\Delta^{\hat{\otimes}2})_1 \right), \tag{2.223}$$

$$(E^* \hat{\otimes} \Delta^{\hat{\otimes}2})_1 \cong \left(E^1 \otimes (\Delta^{\hat{\otimes}2})_0\right) \oplus \left(E^1 \otimes (\Delta^{\hat{\otimes}2})_0\right), \qquad (2.224)$$

and $(x\cdot)$ means Clifford multiplication whenever x is seen as an element in $C_{0,4}$.

We need to see that ψ induces an isomorphism once restricted to $S^{0,4} \times X$ and $B^{0,4} \times A$. Whenever $x \neq 0$ the homomorphism ψ_x is of full rank, and this implies that ψ restricted to $S^{0,4} \times X$ is an isomorphism. On the other hand, the equation $\psi_x = 0$ cannot be solved whenever the homomorphisms φ and $\overline{\varphi}$ are isomorphisms, and this implies that ψ is an isomorphism over $B^{0,4} \times A$.

Remark 2.9 The explicit definition of the morphism γ in terms of ψ_x of Eqs. (2.221) and (2.222) is essentially due to Atiyah and Segal [2, middle of p.17]. It was originally written as the associated map from quaternionic K-theory to real K-theory shifting by 4 the degree. Here we write it in terms of the Clifford algebra $C_{0,4}$ and therefore generalize it for magnetic equivariant K-theory.

Theorem 2.7 (Degree shift isomorphism using Clifford modules) *The isomorphism β of Proposition 2.7, the degree shift homomorphism γ of Definition 2.13, and the Atiyah-Bott-Shapiro homomorphism α of Eq. (2.142) make the following diagram commutative*

$$
\begin{array}{ccccc}
\widehat{M}^{0,q+1}(G,\phi) & \xrightarrow{\ \text{res}\ } & \widehat{M}^{0,q}(G,\phi) & \xrightarrow{\ \alpha\ } & K_G^{-q}(*) \\
\cong \downarrow \beta & & \cong \downarrow \beta & & \cong \downarrow \gamma \\
\widehat{{}^{\widehat{G}}M}^{0,q+5}(G,\phi) & \xrightarrow{\ \text{res}\ } & \widehat{{}^{\widehat{G}}M}^{0,q+4}(G,\phi) & \xrightarrow{\ \alpha\ } & {}^{\widehat{G}}K_G^{-q-4}(*)
\end{array}
$$

$$(2.225)$$

thus making the degree shift homomorphism γ into an isomorphism. In particular

$$(\gamma \circ \gamma) : K_G^{-q}(*) \xrightarrow{\ \cong\ } K_G^{-q-8}(*) \qquad (2.226)$$

is the 8-periodicity of the magnetic equivariant K-theory.

Proof The homomorphism β is the graded tensor product with $\Delta^{\hat{\otimes}2}$ and therefore it commutes with the restriction maps. Now, consider a $C_{0,q}[G,\phi]$-module $M = (M_0, M_1)$ and let us first calculate $\alpha(\beta(M)) = \alpha(M \hat{\otimes} \Delta^{\hat{\otimes}2})$. This is the triple

$$\left(B^{0,q+4} \times \left(M_0 \times (\Delta^{\hat{\otimes}2})_0 \oplus M_1 \times (\Delta^{\hat{\otimes}2})_1\right), B^{0,q+4} \times \left(M_1 \times (\Delta^{\hat{\otimes}2})_0 \oplus M_1 \times (\Delta^{\hat{\otimes}2})_0\right), \sigma\right)$$

$$(2.227)$$

where

$$\sigma((x,y),(m_0 \otimes \xi_0)) = ((x,y),(ym_0 \otimes \xi_0 + m_0 \otimes x\xi_1)) \qquad (2.228)$$

$$\sigma((x,y),(m_1 \otimes \xi_1)) = ((x,y),(ym_1 \otimes \xi_1 - m_1 \otimes x\xi_1)) \qquad (2.229)$$

where an element in $B^{0,q+4}$ is written as the pair (x, y) with x being the first 4 coordinates and y the last q coordinates, $m_i \in M_i$ and $\xi \in (\Delta^{\hat{\otimes}2})_i$.

On the other hand, calculating $\gamma(\alpha(M))$ we obtain the following. $\alpha(M)$ is the triple

$$\left(B^{0,q} \times M_0, \, B^{0,q} \times M_1, \, \varphi\right) \tag{2.230}$$

with $\varphi(y, m_0) = (y, ym_0)$. Now, in order to define $\gamma(\alpha(M))$ we need a map $\overline{\varphi}$ which is a local inverse of φ. In this particular case we can take

$$\overline{\varphi} : B^{0,q} \times M_1 \rightarrow B^{0,q} \times M_0 \tag{2.231}$$

$$(y, m_1) \mapsto (y, -ym_1) \tag{2.232}$$

since $y^2 = -1$. Therefore $\gamma(\alpha(M))$ is the triple

$$\left(B^{0,4} \times B^{0,q} \times \left(M_0 \times (\Delta^{\hat{\otimes}2})_0 \oplus M_1 \times (\Delta^{\hat{\otimes}2})_1\right), \, B^{0,4} \times B^{0,q} \times \left(M_1 \times (\Delta^{\hat{\otimes}2})_0 \oplus M_0 \times (\Delta^{\hat{\otimes}2})_1\right), \, \eta\right) \tag{2.233}$$

where

$$\eta(x, (y, m_0) \otimes \xi_0)) = (x, \varphi(y, m_0) \otimes \xi_0 + (y, m_0) \otimes x\xi_0) \tag{2.234}$$

$$= (x, (y, ym_0) \otimes \xi_0 + (y, m_0) \otimes x\xi_0) \tag{2.235}$$

$$\eta(x, (y, m_1) \otimes \xi_1)) = (x, -\overline{\varphi}(y, m_1) \otimes \xi_1 - (y, m_1) \otimes x\xi_1) \tag{2.236}$$

$$= (x, (y, ym_1) \otimes \xi_1 - (y, m_1) \otimes x\xi_1). \tag{2.237}$$

$$\square$$

The maps σ and η agree on $B^{0,q+4} \subset B^{0,4} \times B^{0,q}$ and therefore we have the equality $\gamma(\alpha(M)) = \alpha(\beta(M))$.

Applying γ twice, we obtain the 8-periodicity since $(\beta \circ \beta)$ is an isomorphism at the level of isomorphism classes of Clifford modules.

2.9 Complexes of Vector Bundles and Thom Isomorphism

In this section, we elaborate on another familiar construction of the groups $\mathbf{K}_G(X, Y)$, this time in terms of complexes of (G, ϕ)-vector bundles. This formulation is crucial for the computation of the coefficients $\mathbf{K}_G^*$ and for proving the Thom isomorphism. The construction and proof align with the corresponding results for $KU_{G_0}(X)$, as detailed in [13, Sect. 3 and Appendix]. The antilinear structure is naturally incorporated into the complex framework, following the same path.

Definition 2.14 Let (X, A) be a G-pair of locally compact spaces. The category $C_G(X, A)$ of **complexes on X acyclic on A** has

- Objects: sequences

$$E^* : \ldots \xrightarrow{d} E^{i-1} \xrightarrow{d} E^i \xrightarrow{d} E^{i+1} \xrightarrow{d} \ldots \tag{2.238}$$

of (G, ϕ)-vector bundles on X such that $E^i = 0$ when $|i|$ is large, homomorphisms d such that $d \circ d = 0$, and

$$\mathrm{supp}(E^*) := \{x \in X \mid \text{the sequence of vector spaces } E_x^* \text{ is not exact}\} \tag{2.239}$$

is a compact subset of $X - A$.

- Morphisms: $f : E^* \longrightarrow F^*$ are sequences of morphisms $f^i : E^i \longrightarrow F^i$ such that $fd = df$.

The direct sum of vector bundles gives a natural sum of complexes

$$(E^* \oplus F^*)^n = E^n \oplus F^n$$

given by the direct sum of vector bundles on each level.

Definition 2.15 Denote the set of isomorphism classes of complexes on X acyclic on A by $\mathbb{L}_G(X, A)$. This is naturally a semigroup under the sum of complexes.

- Two elements E^*, $F^* \in \mathbb{L}_G(X, A)$ are **homotopic**, $E^* \simeq F^*$, if there is an element $H^* \in \mathbb{L}_G(X \times I, A \times I)$ such that $E^* = H^*|_{X \times \{0\}}$ and $F^* = H^*|_{X \times \{1\}}$.
- Two elements E^*, $F^* \in \mathbb{L}_G(X, A)$ are **equivalent**, $E^* \sim F^*$, if there are elements F_0^*, $F_1^* \in \mathbb{L}_G(X, X)$ such that $E^* \oplus F_0^* \simeq F^* \oplus F_1^*$.

Theorem 2.8 *There is an isomorphism*

$$\mathbb{L}_G(X, A)/\!\sim \xrightarrow{\cong} \mathbf{K}_G(X, A) \tag{2.240}$$

Proof The proof follows that of [13, Proposition 3.1]. The antilinear action of the magnetic groups do not play an important role. When X is compact and $A = \emptyset$, then the isomorphism is simply $E^* \longmapsto \sum_k (-1)^k E^k$. $\square$

Now we are going to describe the product in terms of complexes.

Definition 2.16 The product of two complexes E^*, F^* on X acyclic on A is

$$(E^* \otimes F^*)^n = \oplus_{i+j=n} E^i \otimes F^j. \tag{2.241}$$

We have $\mathrm{supp}(E^* \otimes F^*) = \mathrm{supp}(E^*) \cap \mathrm{supp}(F^*)$, see [13, p. 139]. This defines a product

$$\mathbb{L}_G(X, A) \otimes \mathbb{L}_G(X, B) \longrightarrow \mathbb{L}_G(X, A \cup B). \tag{2.242}$$

If $A = B = \emptyset$, this induces the usual product in $\mathbf{K}_G(X)$.

The product in $\mathbf{K}_G(X)$ extends to $\mathbf{K}_G^*(X)$ making it a graded ring. If $\xi_i \in \mathbf{K}_G^{-p_i}(X)$ (for $i = 1, 2$) is represented by a complex $E_i^* \in \mathbb{L}_G(X \times \mathbb{R}^{p_i})$ with compact support, then the product $\xi_1 \cdot \xi_2 \in \mathbf{K}_G^{-p_1-p_2}(X)$, is represented by the complex $pr_1^* E_1^* \otimes pr_2^* E_2^*$ on $X \times \mathbb{R}^{-p_1-p_2}$, where $pr_i : X \times \mathbb{R}^{p_1} \times \mathbb{R}^{p_2} \longrightarrow X \times \mathbb{R}^{p_i}$.

Now we define the Koszul complex, the key construction to define the Thom isomorphism.

Definition 2.17 Let E be a (G, ϕ)-vector bundle on X and s a G-section of E one can form the **Koszul complex**

$$\cdots \longrightarrow 0 \longrightarrow \mathbb{C} \xrightarrow{d} \Lambda^1 E \xrightarrow{d} \Lambda^2 E \xrightarrow{d} \cdots \tag{2.243}$$

where d is defined by $d(\xi) = \xi \wedge s(x)$ if $\xi \in \Lambda^i E_x$.

This complex is acyclic at all points x for which $s(x) \neq 0$. Let $x \in X$ and suppose $e_1, e_2, \ldots, e_m := s(x) \in E_x$ is a basis. If $v = \sum_{I=(i_1,i_2,\ldots,i_k)} \alpha_I e_{i_1} \wedge e_{i_2} \wedge \ldots \wedge e_{i_k} \in \Lambda^k E_x$ is such that $v \wedge s(x) = 0$, we have

$$0 = \left(\sum_{I=(i_1,i_2,\ldots,i_k)} \alpha_I e_{i_1} \wedge e_{i_2} \wedge \ldots \wedge e_{i_k} \right) \wedge e_m \tag{2.244}$$

$$= \sum_{I=(i_1,i_2,\ldots,i_k)} \alpha_I e_{i_1} \wedge e_{i_2} \wedge \ldots \wedge e_{i_k} \wedge e_m \tag{2.245}$$

$$= \sum_{I=(i_1,i_2,\ldots,i_k),i_K \neq m} \alpha_I e_{i_1} \wedge e_{i_2} \wedge \ldots \wedge e_{I_k} \wedge e_m. \tag{2.246}$$

So $\alpha_I = 0$ for every $I = (i_1, i_2, \ldots, i_k)$ such that $I_k \neq m$, that is

$$v = \sum_{I=(i_1,i_2,\ldots,m)} \alpha_I e_{i_1} \wedge e_{i_2} \wedge \ldots \wedge e_m \in d(\Lambda^{k-1} E_x). \tag{2.247}$$

The most important example of a Koszul complex in our case is the following:

Example 2.10 Let $p : E \longrightarrow X$ be a (G, ϕ)-vector bundle, the pull-back $p^* E$ on E has a natural section which is the diagonal map $\Delta : E \longrightarrow E \times_X E = p^* E$

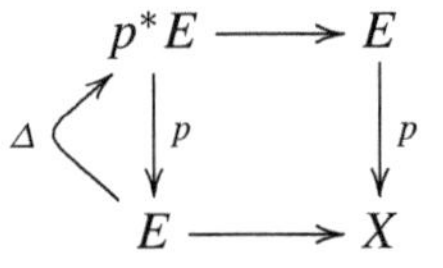

$$\tag{2.248}$$

This section Δ vanishes on the zero-section of E. Let us denote by Λ_E^* the Koszul complex of $p^* E$ and Δ. If X is compact, $\Lambda_E^* \in \mathbf{K}_G(E)$ has compact support and is called the **Thom class**.

Example 2.11 Let V be a representation of (G, ϕ). The exterior algebra $\Lambda^* V$ of V induces a Koszul complex λ_V over V:

$$\lambda_V : \ V \times \{0\} \xrightarrow{\ d\ } V \times \mathbb{C} \xrightarrow{\ d\ } V \times \Lambda^1 V \xrightarrow{\ d\ } V \times \Lambda^2 V \xrightarrow{\ d\ } \dots$$

$$(2.249)$$

where $d(v, w) = v \wedge w$. Of course, this complex is exact on $V \setminus \{0\}$.

Definition 2.18 Let $p : E \longrightarrow X$ be a (G, ϕ)-vector bundle over a compact G-space X. If F^* is a complex with compact support on X then $p^* F^*$ is a complex on E with support $p^{-1}(\mathrm{supp}(F^*))$, and $\Lambda_E^* \otimes p^* F^*$ is a complex with compact support on E. The homomorphism

$$\varphi_* : \mathbf{K}_G(X) \longrightarrow \mathbf{K}_G(E)$$
$$F^* \longmapsto \Lambda_E^* \otimes p^* F^* \qquad\qquad (2.250)$$

is called the **Thom homomorphism**.

One can replace X by $X \times \mathbb{R}^p$ and E by $E \times \mathbb{R}^p$ and induce a Thom homomorphism

$$\varphi_* : \mathbf{K}_G^{-p}(X) \longrightarrow \mathbf{K}_G^{-p}(E).$$

The main theorem of this section is the following:

Theorem 2.9 (The Thom isomorphism) *The Thom isomorphism*

$$\varphi_* : \mathbf{K}_G^{-p}(X) \longrightarrow \mathbf{K}_G^{-p}(E) \qquad\qquad (2.251)$$

is an isomorphism for any (G, ϕ)*-vector bundle* E *on a compact* G*-space* X.

The proof is based on the one given in [25], which uses elliptic differential operators. The guideline of the proof is the following. For every representation V of (G, ϕ):

- Construct a morphism $\alpha : \mathbf{K}_G^*(X \times V) \longrightarrow \mathbf{K}_G^*(X)$ which will be the inverse of the Thom homomorphism $- \otimes \lambda_{V^*} : \mathbf{K}_G^*(X) \longrightarrow \mathbf{K}_G^*(X \times V)$. The isomorphism λ_{V^*} generalizes the Bott isomorphism given in Theorem 2.3.
- In the previous isomorphism replace (G, ϕ) by $(\mathrm{U}(n) \times G, \phi)$, X by $\mathcal{F}(E)$, the frame bundle associated to E, and $V = \mathbb{C}^{\dim E}$ to obtain the Thom isomorphism $\mathbf{K}_G(X) \longrightarrow \mathbf{K}_G(E)$.

The complex, equivariant and Real cases follow these lines, see [25, Sects. 3, 4 and 5].

Remark 2.10 We adopt the proof of [25], based on differential operators and their indexes, for two principal reasons:

- It gives the most general form of the theorem.
- The proofs of the Thom isomorphism given in [1, 13, 14] assume *every vector bundle is locally the sum of line vector bundles. If G is an abelian compact Lie group, then every G-equivariant bundle is locally the sum of line bundles.* This assumption is not true for (G, ϕ)-vector bundles because abelian magnetic groups (G, ϕ) could have complex or quaternion type irreducible representations which are of complex dimension two; for example $(\mathbb{Z}/4, mod_2)$.

2.10   Generalized Bott Isomorphism

We now construct an inverse for $\lambda_{V*} : \mathbf{K}_G^*(X) \longrightarrow \mathbf{K}_G^*(X \times V)$ using the following factorization

$$\mathbf{K}_G(X \times V) \xrightarrow{\quad j \quad} \mathbf{K}_G(X \times \mathrm{P}(V \oplus \mathbf{1})) \xrightarrow{\quad \mathrm{index}(D) \quad} \mathbf{K}_G(X) \tag{2.252}$$

with the map α the composite.

where $\mathbf{1}$ is $\mathbb{C}$ as a magnetic (G, ϕ)-representation on which G acts through ϕ via conjugation and $\mathrm{index}(D_Q)$ is a parametrized index for a specific vector bundle.

Let us describe the first map j and its value at λ_V. Note we have a natural map

$$\mathrm{P}(V \oplus \mathbf{1}) \longrightarrow V^+ \tag{2.253}$$

$$[v, z] \longmapsto \begin{cases} v/z & \text{if } z \neq 0 \\ + & \text{if } z = 0. \end{cases} \tag{2.254}$$

It descends to a homeomorphism

$$\mathcal{H} : \mathrm{P}(V \oplus \mathbf{1})/\mathrm{P}(V) \longrightarrow V^+ \tag{2.255}$$

so we have a homomorphism

$$\mathbf{K}_G(V) \xrightarrow{\mathcal{H}^*} \mathbf{K}_G(\mathrm{P}(V \oplus \mathbf{1}) - \mathrm{P}(V)) \cong \mathbf{K}_G(\mathrm{P}(V \oplus \mathbf{1}), \mathrm{P}(V)) \longrightarrow \mathbf{K}_G(\mathrm{P}(V \oplus \mathbf{1})).$$

with the composite the map j.

$$\tag{2.256}$$

Let us consider the following commutative diagram

$$V^+ \times \Lambda^k V \longleftarrow \Lambda^k(\mathbf{1} \otimes V) \longleftarrow \Lambda^k(H \otimes V) \qquad (2.257)$$

$$V^+ \xleftarrow{\ \mathcal{H}\ } \mathrm{P}(V \oplus \mathbf{1})/\mathrm{P}(V)) \xleftarrow{\ q\ } \mathrm{P}(V \oplus \mathbf{1})$$

where H is the dual of the tautological line bundle on $\mathrm{P}(V \oplus \mathbf{1})$. This is a pull-back diagram. More is true (see [25, Eqs. 4.2 and 4.6]), it induces a diagram of Koszul complexes on each space, so we get

$$j(\lambda_{V^*}) = \sum_i (-1)^i H^i \Lambda^i(V^*), \qquad (2.258)$$

with the very important property that for $Q := j(\lambda_{V^*})$ we have

$$\mathrm{index}(D_Q) = \mathbf{1} \in \mathbf{K}_G(*). \qquad (2.259)$$

Theorem 2.10 *For any compact G-space X and any representation V of (G, ϕ), multiplication by λ_{V^*} induces an isomorphism*

$$\mathbf{K}_G(X) \longrightarrow \mathbf{K}_G(X \times V). \qquad (2.260)$$

***Proof* (Sketch of proof)** As said before, we are going to construct an inverse α. By the previous comments we just focus on the construction of the homomorphism $\mathrm{index}(D)$. This sketch of the proof simply carries out the same procedure presented in [25, Theorem 5.1] where it is shown for magnetic groups of the form $(G_0 \rtimes \mathbb{Z}/2, \pi_2)$ (such as the one in Example 2.7).

As its name suggests, $\mathrm{index}(D)$ is the index of some Fredholm operator. The construction is a generalization of the one given in [25, Sects. 2 and 3]. A more detailed account of this construction can be found in [26, Sect. 1.5] where the analytical subtleties that arise when working with pseudo-differential operators are resolved. One of the important features is the use of the parametrix [26, Sect. 1.3.7].

Consider the Dolbeault complex on the smooth manifold $\mathrm{P} := \mathrm{P}(V \oplus \mathbf{1})$

$$0 \longrightarrow \Omega^{0,0}(\mathrm{P}) \xrightarrow{\ \bar{\partial}\ } \Omega^{0,1}(\mathrm{P}) \xrightarrow{\ \bar{\partial}\ } \Omega^{0,2}(\mathrm{P}) \xrightarrow{\ \bar{\partial}\ } \Omega^{0,3}(\mathrm{P}) \xrightarrow{\ \bar{\partial}\ } \cdots$$

$$(2.261)$$

where $\Omega^{a,b}(\mathrm{P})$ is the complex vector space of (a, b)-differential forms, i.e. sections of the complex vector bundle $\Lambda^a T^*\mathrm{P} \otimes \overline{\Lambda^b T^*\mathrm{P}}$ (see [27, p. 23] or [26, Sect. 1.5.5]).

The action of (G, ϕ) on $V \oplus \mathbf{1}$ induces a smooth action by diffeomorphisms on P. So we have an action on $\Omega^{0,b}(\mathrm{P})$, the sections of the (G, ϕ)-vector bundle of $(0, b)$-forms. Of course, the action commutes with $\bar{\partial}$. Choose an hermitian metric on P. By Lemma 2.9 there is an hermitian (G, ϕ)-invariant metric on $\Omega^{0,b}(\mathrm{P})$, so we can define the formal adjoint operator

$$\bar{\partial}^* : \Omega^{0,b}(\mathrm{P}) \longrightarrow \Omega^{0,b-1}(\mathrm{P}) \tag{2.262}$$

of $\bar{\partial}$ which also commutes with (G, ϕ); for a detailed construction see [28, p.76].

Denote by $\Omega^+(\mathrm{P}) = \oplus_k \Omega^{0,2k}(\mathrm{P})$ and $\Omega^-(\mathrm{P}) = \oplus_k \Omega^{0,2k+1}(\mathrm{P})$ and define the Dirac operator

$$D = \bar{\partial} + \bar{\partial}^* : \Omega^+(\mathrm{P}) \longrightarrow \Omega^-(\mathrm{P}). \tag{2.263}$$

Define the Laplace-Beltrami operator $\Box = \bar{\partial}\bar{\partial}^* + \bar{\partial}^*\bar{\partial} = D^2$. This operator is important for two reasons. First, it allows us to prove that D has a well-defined index. Second, it provides a Hodge decomposition of the Dolbeault complex. The subspace of elements $\omega \in \Omega^{0,b}(\mathrm{P})$ such that $\Box\omega = 0$ will be denoted by $B^{0,b}(\mathrm{P})$ and is called the space of (G, ϕ)-complex harmonic forms. It is easy to show that $\Box\omega = 0$ if and only if $\bar{\partial}\omega = 0 = \bar{\partial}^*\omega$.

By standard arguments, the differential operator $\Box = D^2$ is elliptic and hence D is elliptic too (see [28, Sect. 5] or [26, Sect. 1.3.10]). So we have a well-defined index

$$\mathrm{index}(D) := \ker D - \mathrm{coker}\, D \in \mathbf{R}(G) = \mathbf{K}_G(*). \tag{2.264}$$

We have a Hodge decomposition (see [29, Theorem 15.4.2] or [26, Theorem 1.3.4])

$$\Omega^{0,b}(\mathrm{P}) = \bar{\partial}\Omega^{0,b-1}(\mathrm{P}) \oplus \bar{\partial}^*\Omega^{0,b+1}(\mathrm{P}) \oplus B^{0,b}(\mathrm{P}). \tag{2.265}$$

and from this decomposition we have

$$\mathrm{index}(D) \cong \oplus_k H^{2k}(\mathrm{P}; \mathbb{C}) - \oplus_m H^{2m+1}(\mathrm{P}; \mathbb{C}). \tag{2.266}$$

Now if Q is other (G, ϕ)-vector bundle over P, then D has a natural extension D_Q over the Dolbeaut complex with coefficients in Q and we have a natural (G, ϕ)-isomorphism

$$\ker D_Q \cong \oplus_k H^{2k}(\mathrm{P}; O(Q)) \tag{2.267}$$

$$\mathrm{coker}\, D_Q \cong \oplus_k H^{2k+1}(\mathrm{P}; O(Q)). \tag{2.268}$$

Finally if Q is a (G, ϕ)-vector bundle over $\mathrm{P} \times X$, then we can define the index for the restriction $Q|_{\mathrm{P}\times\{x\}}$ of (G, ϕ)-vector bundles $\forall x \in X$ and produce an element

$$\mathrm{index}(D_Q) \in \mathbf{K}_G(X). \tag{2.269}$$

This $\mathrm{index}(D_Q)$ is the content of the famous Atiyah-Singer index theorem for families. Its construction as well as its properties go beyond the scope of this work. We recommend the original paper of Atiyah and Singer [30] as well as the more recent versions of the theorem that appear in [24, 31] for its construction and properties.

Taking $Q := j(\lambda_{V^*})$, from the Atiyah-Singer index theorem for families we have a homomorphism

$$\text{index}(D_Q) : \mathbf{K}_G(X \times P) \longrightarrow \mathbf{K}_G(X) \tag{2.270}$$

whose image is $\mathbf{1} \in \mathbf{K}_G(*)$ once restricted to points in X (this is shown in [25, Eq. 4.6]). The isomorphism follows from Mayer-Vietoris and the five lemma. $\square$

Proof (**of Theorem** 2.9) We can reduce the case when $p = 0$. Let us recall the classical constructions of the frame and associated bundles.

Let $p : E \longrightarrow X$ be a (G, ϕ)-vector bundle of rank n. Let $x \in X$ and consider the set of linear or antilinear isomorphisms

$$\text{MGL}(\mathbb{C}^n, E_x). \tag{2.271}$$

It has a left action of $(G \times \text{U}(n), \phi)$ given by

$$(g, h) \cdot (f)(z) = gf(h^{-1}z). \tag{2.272}$$

The **frame bundle** is

$$\mathcal{F}(E) := \bigsqcup_x \text{MGL}(\mathbb{C}^n, E_x) \tag{2.273}$$
$$\downarrow$$
$$X.$$

It is a principal fiber bundle over X with structure group $\text{GL}(n, \mathbb{C})$ and an extra action of the magnetic point group (G, ϕ), see for example [32, Sects. 7 and 8]. The action of the normal subgroup $\text{U}(n)$ on $\mathcal{F}(E)$ is free and we have an (G, ϕ)-isomorphism

$$\Psi : \mathcal{F}(E) \times_{\text{U}(n)} \mathbb{C}^n \longrightarrow E$$
$$[(f, z)] \longmapsto f(z). \tag{2.274}$$

This isomorphism is (G, ϕ)-equivariant, indeed

$$g\Psi([f, z]) = g \cdot f(z) \tag{2.275}$$
$$= ((g, 1) \cdot f)(z) \tag{2.276}$$
$$= \Psi(g \cdot [f, z]). \tag{2.277}$$

Thus we have the following commutative diagram of isomorphisms

$$\begin{array}{ccc} \mathbf{K}_{G\times \mathrm{U}(n)}(\mathcal{F}(E)) & \xrightarrow{\ \cong\ } & \mathbf{K}_{G\times \mathrm{U}(n)}(\mathcal{F}(E)\times \mathbb{C}^n) \\ {\scriptstyle\cong}\big\uparrow & & {\scriptstyle\cong}\big\uparrow \\ \mathbf{K}_G(X) & \dashrightarrow[\cong] & \mathbf{K}_G(E) \end{array} \qquad (2.278)$$

where the vertical arrows are given by Theorem 2.1 and the top arrow is given by Theorem 2.10. The Thom isomorphism of Theorem 2.9

$$\varphi_* : \mathbf{K}_G^*(X) \longrightarrow \mathbf{K}_G^*(E) \qquad (2.279)$$

then follows from the lower horizontal isomorphism of diagram (2.278). $\qquad\square$

2.11   Atiyah-Hirzebruch Spectral Sequence

In this section we describe the Atiyah-Hirzebruch spectral sequence (AHSS) for a G-CW complex X. This is an important tool for computing the groups $\mathbf{K}_G^*(X)$ from equivariant Bredon cohomology with coefficients in $\mathbf{K}_G^*$. The following definition is basically due to Bredon [15].

Definition 2.19 Let (G, ϕ) be a finite magnetic group. A G**-CW-complex** is a CW-complex X of finite type (finite number of cells on each dimension) with an action of G by cellular maps such that

$$\{x \in X \mid gx = x\} \qquad (2.280)$$

is a subcomplex of X for every $g \in G$.

A more concrete description is the following: a G-CW-complex X is the union of G-subspaces $X^{(n)}$ such that

- $X^{(0)}$ is a disjoint union of a finite number of orbits G/H and
- $X^{(n+1)}$ is obtained from $X^{(n)}$ by attaching a finite number of G-cells $G/H \times D^{n+1}$ along attaching G-maps $G/H \times S^n \longrightarrow X^{(n)}$.

The attaching maps are determined by their restrictions $S^n \longrightarrow (X^{(n)})^H$. Denote by $\mathrm{Map}_G(X, Y)$ the set of G-equivariant maps from X to Y and by X^H the subspace of poins fixed by $H \leq G$. Then

$$\mathrm{Map}_G(G/H \times S^n, X^{(n)}) \longrightarrow \mathrm{Map}\left(S^n, \left(X^{(n)}\right)^H\right) \qquad (2.281)$$

$$f \longmapsto \widehat{f}(x) := f(H, x) \qquad (2.282)$$

and

$$\mathrm{Map}\left(S^n, \left(X^{(n)}\right)^H\right) \longrightarrow \mathrm{Map}_G(G/H \times S^n, X^{(n)}) \tag{2.283}$$

$$f \longmapsto \widetilde{f}(gH, x) := g \cdot f(x) \tag{2.284}$$

are inverses of each other.

Thus we have

$$X^{(0)} = \sqcup_i G/H_{D_i^0} \tag{2.285}$$

and

$$
\begin{array}{ccc}
\sqcup_j G/H_{D_j^n} \times S_j^{n-1} & \longrightarrow & X^{(n-1)} \\
\downarrow & & \downarrow \\
G/H_{D_j^n} \times D_j^n & \longrightarrow & X^{(n)}.
\end{array}
\tag{2.286}
$$

Now we show that the groups $\widetilde{\mathbf{K}}_G^*(X)$ form a G-cohomology theory in the sense of Matumoto [33].

Definition 2.20 A **reduced G-cohomology theory** on the category of G-finite G-CW complexes with base point and base point preserving G-homotopy classes of base point preserving G-maps is a sequence of contravariant functors $\{\widetilde{h}_G^n\}_{n \in \mathbb{Z}}$ into the category of abelian groups, together with natural transformations

$$\partial^n : \widetilde{h}_G^n(A) \longrightarrow \widetilde{h}_G^{n+1}(X, A) \tag{2.287}$$

for every G-pair (X, A) satisfying the following axioms

- The inclusion $(X, X \cap Y) \to (X \cup Y, Y)$ induces an isomorphism

$$\widetilde{h}_G^n(X \cup Y, Y) \xrightarrow{\cong} \widetilde{h}_G^n(X, X \cap Y) \tag{2.288}$$

- If (X, A) is a pair of G-spaces the following long sequence is exact

$$\cdots \longrightarrow \widetilde{h}_G^{n-1}(A) \xrightarrow{\partial^{n-1}} \widetilde{h}_G^n(X, A) \longrightarrow \widetilde{h}_G^n(X) \longrightarrow \widetilde{h}_G^n(A) \xrightarrow{\partial^n} \widetilde{h}_G^{n+1}(X, A) \longrightarrow \cdots. \tag{2.289}$$

Theorem 2.11 *Let (G, ϕ) be a finite magnetic group. The magnetic (G, ϕ)-equivariant K-theory $\widetilde{\mathbf{K}}_G^*$ is a reduced G-cohomology theory.*

Proof The isomorphism for the inclusion follows from the isomorphism $\mathbf{K}_G(X, Y) \cong \mathbf{K}_G(X/Y)$ shown in Eq. (2.70) and the exactness of the long sequence is shown in Lemma 2.11. $\square$

Remark 2.11 As Matumoto points out [33, p. 54], this reduced G-cohomology theory produces a G-**cohomology theory** by $\mathbf{K}_G^n(X, A) := \widetilde{\mathbf{K}}_G^n(X/A)$.

Theorem 2.12 (Mayer-Vietoris long exact sequence) *Let X be a G-CW-complex and A, B two sub G-CW-complexes such that $X = A \cup B$ and $A \cap B$ is also a sub G-CW-complex. Then there exist a long exact sequence*

$$\ldots \longrightarrow \mathbf{K}_G^n(X) \longrightarrow \mathbf{K}_G^n(A) \oplus \mathbf{K}_G^n(B) \longrightarrow \mathbf{K}_G^n(A \cap B) \longrightarrow \mathbf{K}_G^{n+1}(X) \longrightarrow \ldots \tag{2.290}$$

Proof Consider the following commutative diagram

$$\begin{array}{ccccccccc}
\mathbf{K}_G^n(X) & \xrightarrow{a} & \mathbf{K}_G^n(A) & \longrightarrow & \mathbf{K}_G^n(X, A) & \xrightarrow{r} & \mathbf{K}_G^{n+1}(X) & \longrightarrow & \mathbf{K}_G^{n+1}(A) \\
{\scriptstyle b}\big\downarrow & & {\scriptstyle a'}\big\downarrow & & {\scriptstyle k}\big\downarrow{\scriptstyle \cong} & & \big\downarrow & & \big\downarrow \\
\mathbf{K}_G^n(B) & \xrightarrow{b'} & \mathbf{K}_G^n(B \cap A) & \xrightarrow{c} & \mathbf{K}_G^n(B, B \cap A) & \longrightarrow & \mathbf{K}_G^{n+1}(B) & \longrightarrow & \mathbf{K}_G^{n+1}(B \cap A)
\end{array} \tag{2.291}$$

where the horizontal sequences are the long exact sequences of the pairs (X, A) and $(B, B \cap A)$, respectively, and the vertical arrows are induced by the natural inclusions. The middle vertical arrow is an isomorphism because we have the natural homeomorphism $X/A \cong B/B \cap A$. So we have the long sequence

$$\ldots \longrightarrow \mathbf{K}_G^n(X) \xrightarrow{(a,-b)} \mathbf{K}_G^n(A) \oplus \mathbf{K}_G^n(B) \xrightarrow{a'+b'} \mathbf{K}_G^n(A \cap B) \xrightarrow{rk^{-1}c} \mathbf{K}_G^{n+1}(X) \longrightarrow \ldots \tag{2.292}$$

Standard diagram-chasing arguments shows that this sequence is exact. $\square$

Now we start to construct the Atiyah-Hirzebruch Spectral Sequence (AHSS). The original construction is due to the two authors [34]. The equivariant generalization was done by Segal [13, Sect. 5], and several versions of this spectral sequence have followed. We cite [35, Const. 3.48] for a very general construction of this spectral sequence in the realm of equivariant homotopy theory.

The filtration by G-skeletons

$$X^{(0)} \subset X^{(1)} \subset \ldots \subset X^{(n)} \subset \ldots \subset X \tag{2.293}$$

induces the following diagram

$$\begin{array}{ccc}
\mathbf{K}^*_G(X^{(n+1)}, X^{(n)}) & & \mathbf{K}^*_G(X^{(n)}, X^{(n-1)}) \\
\downarrow j^*_{n+1} & & \downarrow j^*_n \\
\end{array}$$

$$\mathbf{K}^*_G(X) \longrightarrow \cdots \xrightarrow{\; i^*_{n+1}\;} \mathbf{K}^*_G(X^{(n+1)}) \xrightarrow{\; i^*_n\;} \mathbf{K}^*_G(X^{(n)}) \xrightarrow{\; i^*_{n-1}\;} \cdots$$

$$\begin{array}{ccc}
\downarrow \partial^{n+1} & & \downarrow \partial^n \\
\mathbf{K}^{*+1}_G(X^{(n+2)}, X^{(n+1)}) & & \mathbf{K}^{*+1}_G(X^{(n+1)}, X^{(n)})
\end{array}$$

$$(2.294)$$

where i^*_n and j^*_n are induced by the inclusions $X^{(n)} \longrightarrow X^{(n+1)}$ and $(X^{(n)}, \emptyset) \longrightarrow (X^{(n)}, X^{(n-1)})$, respectively, and ∂^n is the connecting homomorphism from the long exact sequence of the pair $(X^{(n+1)}, X^{(n)})$.

If we add over n we obtain the exact triangle

$$\oplus_n \mathbf{K}^*_G(X^{(n)}) \xrightarrow{\quad i = \oplus i^*_n \quad} \oplus_n \mathbf{K}^*_G(X^{(n)}) \qquad (2.295)$$

$$j = \oplus_n j^*_n \qquad\qquad k = \oplus_n \partial_n$$

$$E_1 = \oplus_n \mathbf{K}^*_G(X^{(n)}, X^{(n-1)})$$

By standard argument in the construction of spectral sequences, see for example [35, 36], we obtain an spectral sequence $(E_r, d_r)_{r\in\mathbb{N}}$ whose first page is given by:

$$E^{n,t}_1 = \mathbf{K}^{n+t}_G(X^{(n)}, X^{(n-1)}) \tag{2.296}$$

$$\cong \widetilde{\mathbf{K}}^{n+t}_G(X^{(n)}/X^{(n-1)}) \text{ Definition} \tag{2.297}$$

$$\cong \widetilde{\mathbf{K}}^{n+t}_G\left(\bigvee_j \left((G/H_{D^n_j})^+ \wedge S^n_j \right) \right) \text{ Collapse de } n-1 \text{skeleton} \tag{2.298}$$

$$\cong \bigoplus_j \widetilde{\mathbf{K}}^{n+t}_G\left((G/H_{D^n_j})^+ \wedge S^n_j \right) \text{ Wedge axiom} \tag{2.299}$$

$$= \bigoplus_j \widetilde{\mathbf{K}}^{n+t}_G\left(\Sigma^n (G/H_{D^n_j})^+ \right) \text{ Definition of suspension} \tag{2.300}$$

$$\cong \bigoplus_j \mathbf{K}^t_G(G/H_{D^n_j}) \text{ Suspension isomorphism} \tag{2.301}$$

$$\cong \bigoplus_j \begin{cases} \mathbf{K}^t_{H_{D^n_j}}(*) & \text{if } H_{D^n_j} \nleq G_0 \\ KU^t_{H_{D^n_j}}(*) & \text{if } H_{D^n_j} \leq G_0 \end{cases} \text{ Lemma 2.3} \tag{2.302}$$

$$= C^n_G(X; \mathbf{K}^t_G). \tag{2.303}$$

Remark 2.12 Here $C^n_G(X; \mathbf{K}^t_G)$ denotes the n-dimensional G-cochain group of the G-CW-complex X with local coefficients in $\mathbf{K}^t_G$, i.e.

$$C_G^n\left(X; \mathbf{K}_G^t\right) := \bigoplus_j \mathbf{K}_G^t(G/H_{D_j^n}). \tag{2.304}$$

These groups and their coboundary operators were defined by Bredon [15]. In [37] one can find an overview of these cohomology groups together with some important applications. We will describe such coboundary operators following [38, Sect. 2] by dualizing the constructions given there to reproduce the cohomological counterpart. For each integer $n \geq 0$, let the equivariant n-cells of X be indexed by a set B_n. For $b \in B_n$ the corresponding cell is $G/H_b \times D^n$, adjoined along the G-map $f_b : G/H_b \times \partial D^n \longrightarrow X^{(n-1)}$. We define

$$\delta : C_G^n\left(X; \mathbf{K}_G^t\right) = \bigoplus_{b \in B_n} \mathbf{K}_G^t(G/H_{D_b}) \longrightarrow C_G^{n+1}\left(X; \mathbf{K}_G^t\right) = \bigoplus_{c \in B_{n+1}} \mathbf{K}_G^t(G/H_{D_c})$$

$$\tag{2.305}$$

by defining the restrictions

$$\delta_{b,c} : \mathbf{K}_G^t(G/H_b) \longrightarrow \mathbf{K}_G^t(G/H_c) \tag{2.306}$$

for $b \in B_n$ and $c \in B_{n+1}$.

- If $n = 0$, then $D^{n+1} = [0, 1]$: Let the image of $f_c|_{G/H_c \times \{1\}}$ lie in $G/H_{b_1} \times D^0$ and let the image of $f_c|_{G/H_c \times \{0\}}$ lie in $G/H_{b_0} \times D^0$. Then $f_c|_{G/H_c \times \partial D^1}$ induces a G-map $f_{c,i} : G/H_c \longrightarrow G/H_{b_i}$ for $i = 0, 1$. Define for $s \in \mathbf{K}_G^t(G/H_b)$,

$$\delta_{b,c}(s) = \begin{cases} (-1)^{i+1} \mathbf{K}_G^t(f_{c,i})(s) & \text{if } b = b_i \text{ and } b_0 \neq b_1 \\ \mathbf{K}_G^t(f_{c,1})(s) - \mathbf{K}_G^t(f_{c,0})(s) & \text{if } b = b_0 = b_1 \\ 0 & \text{if } b \neq b_0 \text{ and } b \neq b_1. \end{cases} \tag{2.307}$$

- If $n \geq 1$, we observe that the composition

$$k : G/H_c \times D^{n+1} \xrightarrow{f_c} X^{(n)} \xrightarrow{p} (G/H_b \times D^n)/(G/H_b \times \partial D^n) \tag{2.308}$$

is a well-defined G-map where p is the canonical projection. It is easily seen, via Theorem 2.1 of [38], that k may be G-homotoped to a map $\bar{k}$ such that the induced orbit map $\bar{k}/G : \partial D^{n+1} \longrightarrow D^n/\partial D^n$ is transverse regular at $0 \in D^n/\partial D^n$. Hence $\bar{k}^{-1}(G/H_b \times \{0\})$ is $G/H_c \times \{x_1, x_2, \ldots, x_m\}$ for some finite set of points $x_i \in \partial D^{n+1}$. For each $i, i = 1, 2, \ldots, m$, let

$$\epsilon_i = \begin{cases} +1 & \text{if } \bar{k}/G \text{ preserves orientation near } x_i \\ -1 & \text{if } \bar{k}/G \text{ reverses orientation near } x_i. \end{cases} \tag{2.309}$$

Let $k_i : G/H_c \times \{x_i\} \longrightarrow G/H_b \times \{0\}$ be the G-map induced from $\bar{k}$. For $s \in \mathbf{K}_G^t(G/H_b)$ define

$$\delta_{h,t}(s) := \sum_{i=1}^{m} \epsilon_i \mathbf{K}_G^t(k_i)(s).$$
(2.310)

The cohomology of this cochain complex is denoted by

$$H_G^n(X; \mathbf{K}_G^t) := H^n\left((C^*(X; \mathbf{K}_G^t), \delta)\right).$$
(2.311)

It is clear that the first differential $d_1 : E_1^{n,t} \longrightarrow E_1^{n+1,t}$ can be identified with the coboundary operator δ of the CW-complex cochain $\left(C_G^*\left(X; \mathbf{K}_G^t\right), \delta\right)$. Thus the second page of the spectral sequence is isomorphic to

$$E_2^{n,t} \cong H^n\left(X; \mathbf{K}_G^t\right).$$
(2.312)

The filtration by skeletons also induces a filtration of $\mathbf{K}_G^*(X)$ given by

$$F^t\mathbf{K}_G^*(X) := \ker\left(\mathbf{K}_G^*(X) \longrightarrow \mathbf{K}_G^*(X^{(t-1)})\right)$$
(2.313)

satisfying

$$\ldots \subset F^{t+1}\mathbf{K}_G^*(X) \subset F^t\mathbf{K}_G^*(X) \subset \ldots \subset \mathbf{K}_G^*(X).$$
(2.314)

Summarizing, we have the following result.

Theorem 2.13 (Atiyah-Hirzebruch Spectral Sequence) *Let (G, ϕ) be a finite magnetic point group and X a finite G-CW-complex. Then the Atiyah-Hirzebruch spectral sequence,*

$$E_2^{n,t} \cong H^n\left(X; \mathbf{K}_G^t\right) \Longrightarrow \mathbf{K}_G^*(X),$$
(2.315)

is a right half-plane spectral sequence with differentials $d_r : E_r^{n,t} \longrightarrow E_r^{n+r,t-r+1}$ that converges strongly to $\mathbf{K}_G^*(X^n)$.

Proof The convergence means that $E_\infty^{t,*} \cong F^t\mathbf{K}_G^*(X)/F^{t+1}\mathbf{K}_G^*(X)$ and it is strong if the filtration is complete Hausdorff (see [39, Definition 5.2]). We chose X to be a finite G-CW-complex, so both conditions are satisfied.

Whenever X is not a finite G-CW complex, the convergence might be conditional (see [39, Theorem 12.6] for further details). $\qquad\square$

Remark 2.13 For most of the physical applications, X has a finite number of cells, so the spectral sequence converges to $\mathbf{K}_G^*(X)$. Moreover, convergence implies a graded isomorphism between the **associated graded group**

$$Gr\mathbf{K}_G^*(X) := \oplus_t F^t\mathbf{K}_G^*(X)/F^{t+1}\mathbf{K}_G^*(X)$$
(2.316)

and the final page of the spectral sequence, that is, the convergence means that we
have isomorphisms

$$E_\infty^{t,n-t} \cong F^t \mathbf{K}_G^n(X)/F^{t+1}\mathbf{K}_G^n(X) \qquad \text{for all } n \in \mathbb{Z}, t \geq 0. \qquad (2.317)$$

If we want to recover $\mathbf{K}_G^n(X)$ from the associated graded $Gr\mathbf{K}_G^n(X)$ we need to
inductively resolve the extension problems that may arise in the short exact sequences

$$0 \longrightarrow F^{t+1}\mathbf{K}_G^n(X) \longrightarrow F^t\mathbf{K}_G^n(X) \longrightarrow E_\infty^{t,n-t} \longrightarrow 0 \,. \qquad (2.318)$$

In the case that all these short exact sequences split, then we can recover $\mathbf{K}_G^n(X)$
directly from the final page of the spectral sequence as the sum $\oplus_t E_\infty^{t,n-t}$. However,
in general the sequences of diagram (2.318) may not split and we would need to
solve the extension problems inductively. An example of a non-split extension is
presented in the sequence of diagram (3.25) for the case $(G, \phi) = (\mathbb{Z}/4, \mathrm{mod}\ 2)$ and
$X = S^1 \times S^1$.

References

1. Atiyah, M.F.: K-theory and reality. Quart. J. Math. Oxford Ser. **2**(17), 367–386 (1966)
2. Atiyah, M.F., Segal, G.B.: Equivariant K-theory and completion. J. Diff. Geom. **3**, 1–18 (1969)
3. Atiyah, M.F., Segal, G.: Twisted K-theory. Ukr. Mat. Visn. **1**(3), 287–330 (2004)
4. Karoubi, M.: Sur la K-th orie quivariante. In: S minaire Heidelberg-Saarbr cken-Strasbourg sur la K-Th orie, pp. 187–253. Springer, Berlin, Heidelberg (1970)
5. Freed, D.S., Moore, G.W.: Twisted equivariant matter. Ann. Henri Poincaré **14**(8), 1927–2023 (2013)
6. Gomi, K.: Freed-Moore K-theory. Commun. Anal. Geom. **31**(4), 979–1067 (2023)
7. Shiozaki, K., Sato, M., Gomi, K.: Topological crystalline materials: general formulation, module structure, and wallpaper groups. Phys. Rev. B **95**, 235425 (2017)
8. Shiozaki, K., Sato, M., Gomi, K.: Atiyah-hirzebruch spectral sequence in band topology: general formalism and topological invariants for 230 space groups. Phys. Rev. B **106**, 165103 (2022)
9. Kitaev, A.: Periodic table for topological insulators and superconductors. In: AIP Conference Proceedings, vol. 1134, pp. 22–30. American Institute of Physics (2009)
10. Rau, J.G., Kin-Ho Lee, E., Kee, Hae-Young., Iridates and related materials: Spin-orbit physics giving rise to novel phases in correlated systems. Ann. Rev. Conden. Matter Phys. **7**, 195–221 (2016)
11. Hartshorne, R.: Stable vector bundles and instantons. Commun. Math. Phys. **59**(1), 1–15 (1978)
12. Dupont, J.L.: Symplectic bundles and KR-theory. Math. Scand. **24**, 27–30 (1969)
13. Segal, G.: Equivariant K-theory. Inst. Hautes Études Sci. Publ. Math. **34**, 129–151 (1968)
14. Atiyah, M.F.: K-theory. W. A. Benjamin, Inc., New York, Amsterdam. Lecture notes by D. W. Anderson (1967)
15. Bredon, G.E.: Equivariant cohomology theories, volume No. 34 of Lecture Notes in Mathematics. Springer, Berlin, New York (1967)
16. Murayama, M.: On G-ANRs and their G-homotopy types. Osaka J. Math. **20**(3), 479–512 (1983)
17. Palais, R.S.: The Classification of G-spaces. Memoirs of the American Mathematical Society, vol. 36. American Mathematical Society, Providence, RI (1960)

18. Karoubi, M.: K-theory. Classics in Mathematics. Springer, Berlin (2008). An introduction, Reprint of the 1978 edition, With a new postface by the author and a list of errata

19. Atiyah, M.F., Bott, R: On the periodicity theorem for complex vector bundles. Acta Math. **112**, 229–247 (1964)

20. Hoffman, K.: Banach Spaces of Analytic Functions. Dover Publications (2014)

21. Atiyah, M.F., Bott, R., Shapiro, A.: Clifford modules. Topology **3**(suppl), 3–38 (1964)

22. Adem, A., Ruan, Y.: Twisted orbifold K-theory. Commun. Math. Phys. **237**(3), 533–556 (2003)

23. Dwyer, C.: Twisted equivariant K-theory for proper actions of discrete groups. K-Theory **38**(2), 95–111 (2008)

24. Blaine Lawson, Jr. H., Michelsohn, M.-L.: Spin geometry, volume 38 of Princeton Mathematical Series. Princeton University Press, Princeton, NJ (1989)

25. Atiyah, M.F.: Bott periodicity and the index of elliptic operators. Quart. J. Math. Oxford Ser. **2**(19), 113–140 (1968)

26. Gilkey, P.B.: The geometry of spherical space form groups, volume 28 of Series in Pure Mathematics, 2nd edn. World Scientific Publishing Co. Pte. Ltd., Hackensack, NJ (2018)

27. Griffiths, P., Harris, J.: Principles of algebraic geometry. Wiley Classics Library. Wiley Inc., New York (1994). Reprint of the 1978 original

28. Palais, R.S.: Seminar on the Atiyah-Singer index theorem, volume No. 57 of Annals of Mathematics Studies. Princeton University Press, Princeton, NJ. With contributions by Atiyah, M.F., Borel, A., Floyd, E.E., Seeley, R.T., Shih, W., Solovay R. (1965)

29. Hirzebruch, F.: Topological methods in algebraic geometry. Classics in Mathematics. Springer-Verlag, Berlin, english edition, 1995. Translated from the German and Appendix One by Schwarzenberger, R.L.E., Appendix Two by Borel, A., Reprint of the 1978 edition

30. Atiyah, M.F., Singer, I.M.: The index of elliptic operators. IV. Ann. Math. **2**(93), 119–138 (1971)

31. Berline, N., Getzler, E., Vergne, M.: Heat kernels and Dirac operators. Grundlehren Text Editions. Springer, Berlin (2004). Corrected reprint of the 1992 original

32. Husemöller, D.: Fibre bundles, volume 20 of Graduate Texts in Mathematics, 3rd edn. Springer, New York (1994)

33. Matumoto, T.: Equivariant cohomology theories on G-CW complexes. Osaka Math. J. **10**, 51–68 (1973)

34. Atiyah, M.F., Hirzebruch, F.: Vector bundles and homogeneous spaces. In: Proceedings of the Symposia in Pure Mathematics, vol. III, pp. 7–38. American Mathematical Society, Providence, RI (1961)

35. Degrijse, D., Hausmann, M., Luck, W., Patchkoria, I., Schwede, S.: Proper equivariant stable homotopy theory. Mem. Am. Math. Soc. **288**(1432), vi+142 (2023)

36. Kono, A., Tamaki, D.: Generalized cohomology, volume 230 of Translations of Mathematical Monographs. American Mathematical Society, Providence, RI, Japanese edition. Iwanami Series in Modern Mathematics (2006)

37. Bárcenas, N.: A survey of computations of Bredon cohomology. In: Tu, L.W. (ed.) Group Actions and Equivariant Cohomology. Contemporary Mathematics, vol. 808, pp. 1–28. American Mathematical Society, Providence, RI (2024)

38. Willson, S.J.: Equivariant homology theories on G-complexes. Trans. Am. Math. Soc. **212**, 155–171 (1975)

39. Michael Boardman, J.: Conditionally convergent spectral sequences. In: Homotopy invariant algebraic structures (Baltimore, MD, 1998), volume 239 of Contemporary Mathematics, pp. 49–84. American Mathematical Society, Providence, RI (1999)

40. Bárcenas, J.E., Joachim, M., Uribe, B.: Universal twist in equivariant K-theory for proper and discrete actions. Proc. Lond. Math. Soc. (3) **108**(5), 1313–1350 (2014)

Chapter 3
Applications in Topological Phases of Matter

Abstract In this chapter we use the Atiyah-Hirzebruch spectral sequence to determine the whole magnetic equivariant K-theory groups of the two dimensional torus endowed with the rotation action of the cyclic magnetic group of order four. Furthermore, we furthermore show how to calculate the twisted version of this magnetic equivariant K-theory, defined by the group extension of the magnetic cyclic group of order eight.

Keywords Atiyah-Hirzebruch spectral sequence · Spin-orbit coupling · Torus · Magnetic equivariant K-theory · Twisted magnetic equivariant K-theory

Among the many interesting applications of K-theory in physics, one that has shown to be very influential is the incorporation of K-theory groups into the classification of gapped phases of interacting fermions [1, 2]. The ten-fold way classification presented in the periodic table for topological insulators and superconductors by Kitaev [3] after the work of Altland and Zirnbauer [4], paved the way for the immersion of real K-theoretical invariants in the study and classification of crystals [5, 6].

The approach to study gapped interaction of fermions through non-commutative geometric tools à la Connes [7] was brilliantly initiated by Bellissard, van Elst and Schulz–Baldes [1] and further developed by many others (see [8] and the references therein). It was the work of Freed and Moore [6] that organized most of the previous information on K-theory invariants noting that the more general description needed incorporated not only equivariant real K-theory but moreover their twisted versions. Gomi went further in [9], and using the language of groupoids presented the generalization of several properties of the real equivariant K-theory to the more general magnetic case.

The present work arose as attempt to provide an alternative description of the properties of the magnetic equivariant K-theory that is better suited for the applications in condensed matter physics. We hope to have accomplished this task in the previous sections. Now we are going to present some non-trivial calculations of magnetic equivariant K-theory groups that are of relevance in the study of altermagnets [10–12]. We believe that these examples are interesting enough to justify the importance of presenting the properties of the magnetic equivariant K-theory in an

B. Uribe Jongbloed et al., *Magnetic Equivariant K-Theory*,
SpringerBriefs in Mathematics, https://doi.org/10.1007/978-3-032-05914-7_3

organized manner, and moreover, that they are generically enough to allow the reader to be able to mimic our calculations to other similar systems.

Before presenting the calculations we must first describe the symmetries that appear in systems with spin-orbit coupling.

3.1 Symmetries Under Spin-Orbit Interaction

The magnetic groups that we want to consider are called magnetic point groups in condensed matter physics. Moreover, we need to consider magnetic point groups in the spin-orbit interaction setup.

Let us review quickly their definition. Crystals symmetries are three dimensional rigid motions that leave the lattice structure of the crystal fixed; this is the space group of symmetries. The time reversal symmetry $\mathbb{T}$ is always present in these materials.

If the crystal has internal magnetization, the time reversal symmetry is broken, but combinations of time reversal symmetry with rigid motions may preserve the lattice structure of the magnetic material. This group of symmetries is called the magnetic space group.

Using the language of Lie groups we have that the rigid motions are $\mathbb{R}^3 \rtimes O(3)$ and the time reversal symmetry defines a group of order 2. The magnetic space group $\mathcal{G}$ is thus a subgroup of $[\mathbb{R}^3 \rtimes O(3)] \times \mathbb{Z}/2$ that defines the short exact sequence of groups that appear on the second row of the following diagram:

$$
\begin{array}{ccccc}
\mathbb{R}^3 & \longrightarrow & [\mathbb{R}^3 \rtimes O(3)] \times \mathbb{Z}/2 & \longrightarrow & O(3) \times \mathbb{Z}/2 \\
\uparrow & & \uparrow & & \uparrow \\
\Lambda & \longrightarrow & \mathcal{G} & \longrightarrow & G.
\end{array}
\tag{3.1}
$$

Here the copy of $\mathbb{Z}/2$ means the group generated by time reversal symmetry, $\Lambda \cong \mathbb{Z}^3$ is the group of translational symmetries of the crystal and the quotient $\mathcal{G}/\Lambda = G$ is known as the magnetic point group. The map $\phi : G \to \mathbb{Z}/2$ is canonically induced from the composition $G \to O(3) \times \mathbb{Z}/2 \xrightarrow{\pi_2} \mathbb{Z}/2$; this homomorphism separates the elements in G which are built from compositions of time reversal with rigid symmetries from the rigid symmetries alone.

The magnetic $\mathcal{G}$-equivariant vector bundles over $\mathbb{R}^3$ are the ones of interest in condensed matter physics since they represent the occupied states of a gapped Hamiltonian. Therefore the topological invariants coming from magnetic $\mathcal{G}$-equivariant K-theory are important.

The magnetic space group $\mathcal{G}$ acts on $\mathbb{R}^3$ and this action induces an action of the magnetic point group G on $\mathbb{R}^3/\Lambda \cong (S^1)^3$. Whenever the group of crystal symmetries is *symmorphic*, namely that the extension $\mathcal{G} \to G$ splits, then we have an isomorphism between the magnetic $\mathcal{G}$-equivariant K-theory of $\mathbb{R}^3$ and the magnetic G-equivariant K-theory of $(S^1)^3$:

$$\mathbf{K}_G^*(\mathbb{R}^3) \cong \mathbf{K}_G^*((S^1)^3). \tag{3.2}$$

This K-theory group $\mathbf{K}_G^*((S^1)^3)$ provides topological invariants for the structure of the energy bands in a gapped Hamiltonian.

When studying electronic properties of magnetic materials the spin of the electron needs to be taken into account. The appropriate group of point symmetries becomes the double cover $\mathrm{Pin}_-(3)$ of $\mathrm{O}(3)$, which is isomorphic to $\mathrm{SU}(2) \times \mathbb{Z}/2$. Here the inversion symmetry $\mathbf{r} \mapsto -\mathbf{r}$ in $\mathrm{O}(3)$ lifts to a symmetry of order two in $\mathrm{Pin}_-(3)$, and therefore all mirror symmetries in $\mathrm{O}(3)$ lift in $\mathrm{Pin}_-(3)$ to symmetries that square to -1.

The time reversal operator, in the presence of spin orbit interaction, squares to -1, and this -1 must agree with the -1 of the square of the lifts of the mirror symmetries. In general the time reversal operator is taken to be $i\sigma_2\mathbb{K}$ where σ_2 is the second Pauli matrix and $\mathbb{K}$ is complex conjugation. Therefore the group of point symmetries in the presence of spin orbit interaction is the quotient group

$$([\mathrm{SU}(2) \times \mathbb{Z}/4]/\Delta) \times \mathbb{Z}/2 \tag{3.3}$$

where $\Delta := \langle -\mathrm{Id}, 2 \rangle \cong \mathbb{Z}/2$.

If we consider the group $\mathbb{Z}/4$ as the one generated by the operator $i\sigma_2\mathbb{K}$, we see that $(i\sigma_2\mathbb{K})^2 = -\mathrm{Id}$ and therefore it is clear why we need to quotient by the subgroup Δ.

Incorporating the previous information in the following diagram:

$$\mathbb{Z}/2 \longrightarrow ([\mathrm{SU}(2) \times \mathbb{Z}/4]/\Delta) \times \mathbb{Z}/2 \longrightarrow \mathrm{SO}(3) \times \mathbb{Z}/2 \times \mathbb{Z}/2 \tag{3.4}$$

$$\mathbb{Z}/2 \longrightarrow \widetilde{G} \longrightarrow G$$

we see that the group $([\mathrm{SU}(2) \times \mathbb{Z}/4]/\Delta) \times \mathbb{Z}/2$ defines a $\mathbb{Z}/2$-central extension of $\mathrm{O}(3) \times \mathbb{Z}/2$ (here we are using the canonical isomorphism $\mathrm{O}(3) \cong \mathrm{SO}(3) \times \mathbb{Z}/2$) where $\mathbb{Z}/2$ acts by multiplication by -1, and $\widetilde{G}$ is the induced extension over G.

The group $\widetilde{G}$ incorporates the symmetries that interact accordingly with the spin, and therefore this group is the one that will act on the fibers of the magnetic vector bundles. Of particular importance is that the kernel of the homomorphism $\widetilde{G} \to G$ acts on the fibers by the sign representation.

Definition 3.1 Let $G \subset \mathrm{O}(3) \times \mathbb{Z}/2$ be a magnetic point group. Denote by $\widetilde{G}^{soc} := \widetilde{G}$ the $\mathbb{Z}/2$-central extension of G defined in diagram (3.4) and call it the *SOC group* of G.

In the presence of spin orbit interaction, in a crystal whose groups of symmetries is symmorphic, the topological invariant of interest is therefore the $\widetilde{G}^{soc}$-twisted G-equivariant K-theory of $(S^1)^3$.

Definition 3.2 Let $\mathcal{G}$ be a symmorphic space magnetic group and G its magnetic point group. Denote by $^{\widetilde{G}^{soc}}\mathbf{K}_G((S^1)^3)$ the $\widetilde{G}^{soc}$-twisted G-equivariant K-theory groups of the torus $(S^1)^3$, where the kernel of the homomorphism $\widetilde{G}^{soc} \to G$ acts by multiplication by -1.

Particular choices of magnetic point groups and actions are of importance in understanding topological invariants of magnetic crystals. In what follows we will only concentrate in the magnetic groups that are generated by a composition of a four-fold rotation C_4 around the z-axis and time reversal $\mathbb{T}$, in both the spin and no spin orbit interaction, and in $\mathbb{R}^2$.

3.2   2-Dimensional Torus with $C_4\mathbb{T}$ Symmetry

Consider the 2-dimensional torus $T^2 = S^1 \times S^1$ with periodic coordinates (x, y) where the 4-fold rotation action is $C_4(x, y) = (-y, x)$ and the time reversal operator acts by inverting the coordinates $\mathbb{T}(x, y) = (-x, -y)$. The composition of both operations gives the following action:

$$C_4\mathbb{T}(x, y) = (y, -x). \tag{3.5}$$

We will be interested in the magnetic cyclic group

$$G := \langle C_4\mathbb{T} \rangle \cong \mathbb{Z}/4 \tag{3.6}$$

acting in the 2-dimensional torus T^2. We will use the Atiyah-Hirzebruch spectral sequence to calculate both the magnetic equivariant K-theory and the twisted one in the presence of spin orbit interaction. Let us start by showing the $\widetilde{G}$-CW decomposition of the torus T^2.

The $\mathbb{Z}/4$-CW decomposition of the 2-dimensional torus T^2 can be seen in Fig. 3.1 and consists of the following cells:

$$0-\text{cells} : \underbrace{\left(\mathbb{Z}/4 \Big/ \mathbb{Z}/4\right) \times D^0}_{\Gamma} \sqcup \underbrace{\left(\mathbb{Z}/4 \Big/ \mathbb{Z}/2\right) \times D^0}_{A \sqcup A'} \sqcup \underbrace{\left(\mathbb{Z}/4 \Big/ \mathbb{Z}/4\right) \times D^0}_{X}, \tag{3.7}$$

$$1-\text{cells} : \underbrace{(\mathbb{Z}/4) \times D^1}_{\gamma} \sqcup \underbrace{(\mathbb{Z}/4) \times D^1}_{\sigma}, \tag{3.8}$$

$$2-\text{cells} : \underbrace{(\mathbb{Z}/4) \times D^2}_{\Omega}. \tag{3.9}$$

We are interested in calculating the magnetic equivariant K-theory $\mathbf{K}^*_{\mathbb{Z}/4}(T^2)$ in order to find out the topological invariants that materials with a point group symmetry generated by $C_4\mathbb{T}$ may have. Since the magnetic group is cyclic of order bigger than 2, then we know from Proposition 2.6 that the K-theory $\mathbf{K}^*_{\mathbb{Z}/4}(T^2)$ is 4-periodic. Therefore we will only calculate $\mathbf{K}^*_{\mathbb{Z}/4}(T^2)$ for $* = 0, -1, -2, -3$.

Applying the procedure of the Atiyah-Hirzebruch spectral sequence of §2.11, we see that the first page becomes

$$E_1^{0,*} \cong \mathbf{K}_{\mathbb{Z}/4}^*(\{\Gamma\}) \oplus K U_{\mathbb{Z}/2}^*(\{A\}) \oplus \mathbf{K}_{\mathbb{Z}/4}^*(\{X\}), \tag{3.10}$$

$$E_1^{1,*} \cong K U^*(\{\sigma\}) \oplus K U^*(\{\gamma\}), \tag{3.11}$$

$$E_1^{2,*} \cong K U^*(\{\Omega\}). \tag{3.12}$$

In Eq. (2.5) it is shown that $\mathbf{K}_{\mathbb{Z}/4}^* \cong K\mathbb{R}^* \oplus K\mathbb{H}^*$ where $\mathbf{R}(\mathbb{Z}/4) \cong \mathbb{Z} \oplus \mathbb{Z}$ is generated by the irreducible representations defined by complex conjugation $\mathbb{K}$ and the 2×2 matrix $i\sigma_2\mathbb{K}$; denote them by R_1 and R_2 respectively. The K-theory group $K U_{\mathbb{Z}/2}^*$ is 2-periodic with $K U_{\mathbb{Z}/2}^0 \cong \mathbb{Z} \oplus \mathbb{Z}$ generated by the irreducible representations of $\mathbb{Z}/2$ which we denote R_1' and R_2', and $K U^*$ is $\mathbb{Z}$ in even degrees. Hence the first page of the spectral sequence becomes:

$$
\begin{aligned}
E_1^{0,0} = \mathbb{Z}^2 \oplus \mathbb{Z}^2 \oplus \mathbb{Z}^2 \;&\xrightarrow{d_1^{0,0}}\; E_1^{1,0} = \mathbb{Z} \oplus \mathbb{Z} \;\xrightarrow{d_1^{1,0}}\; E_1^{2,0} = \mathbb{Z}, \\
E_1^{0,-1} = \mathbb{Z}/2 \oplus 0 \oplus \mathbb{Z}/2 \;&\xrightarrow{d_1^{0,-1}}\; E_1^{1,-1} = 0 \oplus 0 \;\xrightarrow{d_1^{1,-1}}\; E_1^{2,-1} = 0, \\
E_1^{0,-2} = \mathbb{Z}/2 \oplus \mathbb{Z}^2 \oplus \mathbb{Z}/2 \;&\xrightarrow{d_1^{0,-2}}\; E_1^{1,-2} = \mathbb{Z} \oplus \mathbb{Z} \;\xrightarrow{d_1^{1,-2}}\; E_1^{2,-2} = \mathbb{Z}, \\
E_1^{0,-3} = 0 \oplus 0 \oplus 0 \;&\xrightarrow{d_1^{0,-3}}\; E_1^{1,-3} = 0 \oplus 0 \;\xrightarrow{d_1^{1,-3}}\; E_1^{2,-3} = 0.
\end{aligned}
\tag{3.13}
$$

Now we are going to focus on the differentials $d_1^{n,t}$. We use the orientations given in Fig. 3.1 and we write the differentials in matrix form:

$$
d_1^{0,0} =
\begin{array}{|cc|cc|cc|c|}
\hline
\multicolumn{2}{|c|}{\Gamma} & \multicolumn{2}{c|}{A} & \multicolumn{2}{c|}{X} & \\
R_1 & R_2 & R_1' & R_2' & R_1 & R_2 & \\
\hline
0 & 0 & -1 & -1 & 1 & 2 & \gamma \\
\hline
-1 & -2 & 1 & 1 & 0 & 0 & \upsilon \\
\hline
\end{array}
\qquad
d_1^{0,-2} =
\begin{array}{|cc|c|}
\hline
\multicolumn{2}{|c|}{A} & \\
R_1' & R_2' & \\
\hline
-1 & -1 & \gamma \\
\hline
1 & 1 & \sigma \\
\hline
\end{array}
\tag{3.14}
$$

where $R_1 = \langle \mathbb{K} \rangle$ maps to ± 1, $R_2 = \langle i\sigma_2\mathbb{K} \rangle$ maps to ± 2 and R_i' maps to ± 1 depending whether the vertex is joined to the respective edge.

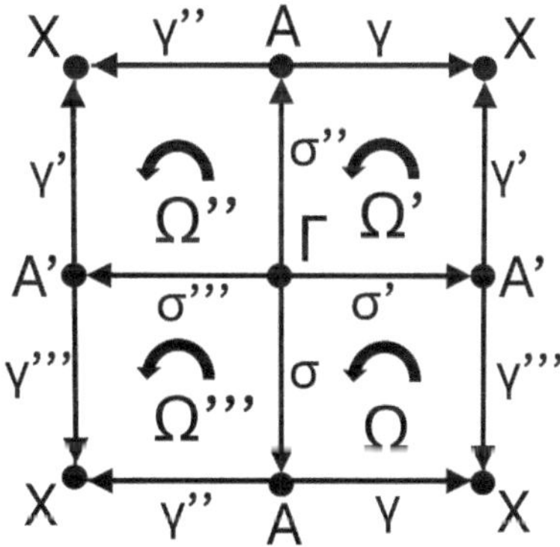

Fig. 3.1 $\mathbb{Z}/4$-CW decomposition of the torus $T^2 = S^1 \times S^1$. The 0-skeleton consists of the points Γ, A, A', X, the 1-skeleton adds the 1-cells generated by the orbits of γ and σ, and the 2-skeleton adds the 2-cell generated by the orbit of Ω

The differentials $d_1^{1,0}$ and $d_1^{1,-2}$ need some extra consideration. The boundary of Ω consists of σ, γ, γ''' and σ' where σ' and γ''' are the paths obtained from σ and γ after the action of $(C_4\mathbb{T})^3$ and $C_4\mathbb{T}$, respectively. The isomorphism induced by the pullback of the generator $C_4\mathbb{T}$

$$(C_4\mathbb{T})^* : KU^*(\gamma''', \partial\gamma''') \xrightarrow{\cong} KU^*(\gamma, \partial\gamma) \tag{3.15}$$

is an isomorphism. More over, it is the composition of the isomorphisms:

$$KU^*(\gamma''', \partial\gamma''') \xrightarrow{(C_4)^*} KU^*(\gamma'', \partial\gamma'') \xrightarrow{\mathbb{T}^*} KU^*(\gamma, \partial\gamma). \tag{3.16}$$

The second isomorphism is the automorphism in KU^* that complex conjugation defines. This is the identity in KU^0 while it is multiplication by -1 in KU^{-2}. In KU^0 the rank of the complex vector bundle is unaffected by complex conjugation, while the Hopf bundle $[H] - \mathbf{1}$ in $KU^{-2} = \widetilde{KU}^0(S^2)$ is mapped to its conjugate bundle $[\overline{H}] - \mathbf{1}$.

This means that $\mathbb{T}^*$ acts trivially on $KU^{-3}(\gamma, \partial\gamma) \cong KU^{-4}$ while it acts by multiplication by -1 on $KU^{-1}(\gamma, \partial\gamma) \cong KU^{-2}$.

Hence, the remaining differentials of the first page are:

$$d_1^{1,0} = \begin{vmatrix} \gamma & \sigma & & \\ 0 & 0 & \Omega \end{vmatrix} \quad d_1^{1,-2} = \begin{vmatrix} \gamma & \sigma & & \\ 2 & 2 & \Omega \end{vmatrix}. \tag{3.17}$$

Therefore, the second page of the spectral sequence becomes:

$$\begin{array}{lll} E_2^{0,0} = \mathbb{Z}^4 & E_2^{1,0} = 0 & E_2^{2,0} = \mathbb{Z} \\ E_2^{0,-1} = (\mathbb{Z}/2)^2 & E_2^{1,-1} = 0 & E_2^{2,-1} = 0 \\ E_2^{0,-2} = (\mathbb{Z}/2)^2 \oplus \mathbb{Z} & E_2^{1,-2} = 0 & E_2^{2,-2} = \mathbb{Z}/2 \\ E_2^{0,-3} = 0 & E_2^{1,-3} = 0 & E_2^{2,-3} = 0. \end{array} \tag{3.18}$$

This implies that all second differentials $d_2^{n,t}$ are zero except for the case $d_2^{0,-1}$: $E_2^{0,-1} \to E_2^{2,-2}$. Let us show that this differential is surjective.

Consider the diagram

$$ \mathbf{K}_{\mathbb{Z}/4}^{-1}(X_1, X_0) \tag{3.19} $$

$$
\begin{array}{ccccccc}
\mathbf{K}_{\mathbb{Z}/4}^{-1}(X_1, X_0) & & & & & & \\
\downarrow{\scriptstyle j_1^*} & \searrow{\scriptstyle d_1^{1,-2}} & & & & & \\
\mathbf{K}_{\mathbb{Z}/4}^{-1}(X_1) & \xrightarrow{\partial^1} & \mathbf{K}_{\mathbb{Z}/4}^{0}(X_2, X_1) & \xrightarrow{j_2^*} & \mathbf{K}_{\mathbb{Z}/4}^{0}(X_2) & \longrightarrow & \mathbf{K}_{\mathbb{Z}/4}^{0}(X_1) \\
\downarrow{\scriptstyle i_0^*} & & & & & & \\
\mathbf{K}_{\mathbb{Z}/4}^{-1}(X_0) & & & & & &
\end{array}
$$

with a horizontal exact sequence, a vertical short exact sequence, and the diagonal homomorphism $d_1^{1,-2}$ being the first differential of the spectral sequence. Here we have followed the notation of diagram (2.294).

The cokernel of the first differential $d_1^{1,-2}$ is the group $\mathbb{Z}/2 = E_2^{2,-2}$ while the image of the second differential $d_2^{0,-1}$ is the quotient group $\frac{\mathrm{Im}(\partial^1)}{\mathrm{Im}(d_1^{1,-2})}$. We claim that this quotient group $\frac{\mathrm{Im}(\partial^1)}{\mathrm{Im}(d_1^{1,-2})}$ is $\mathbb{Z}/2$, thus implying that $d_2^{0,-1}$ is surjective. This claim will follow from showing that the image of the homomorphism j_1^* will not generate the torsion free part of $\mathbf{K}_{\mathbb{Z}/4}^{-1}(X_1)$; let us show this.

Consider the part of the 1-skeleton generated by the orbits of σ and γ; denote them $[\sigma] := \mathbb{Z}/4 \cdot \sigma$ and $[\gamma] := \mathbb{Z}/4 \cdot \gamma$ respectively. These are a $\mathbb{Z}/4$-spaces whose quotient spaces $[\sigma]/(\mathbb{Z}/2)$ $[\gamma]/(\mathbb{Z}/2)$ are both $\mathbb{Z}/2$-homeomorphic to the 1-dimensional ball $B^{1,0}$. Hence the projection map $[\sigma] \to [\sigma]/(\mathbb{Z}/2) \cong B^{1,0}$, together with the excision isomorphisms, induce the following homomorphisms among the short exact sequences of K-theories, which are given by applying Cor. 2.8 to the triples $(B^{1,0}, S^{1,0} \cup \{0\}, S^{1,0})$ and $(X_1, [\gamma] \cup \{\Gamma\}, [\gamma])$:

$$
\begin{array}{ccccc}
\mathbb{Z} \cong K\mathbb{R}^{-1}(B^{1,0}, S^{1,0} \cup \{0\}) & \xrightarrow{\times 2} & \mathbb{Z} \cong K\mathbb{R}^{-1}(B^{1,0}, S^{1,0}) & \longrightarrow & \mathbb{Z}/2 \cong K\mathbb{R}^{-1}(S^{1,0} \cup \{0\}, S^{1,0}) \\
\downarrow{\scriptstyle\cong} & & \downarrow & & \downarrow{\scriptstyle\cong} \\
\mathbb{Z} \cong \mathbf{K}_{\mathbb{Z}/4}^{-1}(X_1, [\gamma] \cup \{\Gamma\}) & \longrightarrow & \mathbf{K}_{\mathbb{Z}/4}^{-1}(X_1, [\gamma]) & \longrightarrow & \mathbb{Z}/2 \cong \mathbf{K}_{\mathbb{Z}/4}^{-1}([\gamma] \cup \{\Gamma\}, [\gamma])
\end{array}
$$

$$(3.20)$$

thus implying that $\mathbf{K}_{\mathbb{Z}/4}^{-1}(X_1, [\gamma]) \cong \mathbb{Z}$ and that the bottom left horizontal map is given by multiplication by 2. Therefore the commutativity of the square

$$
\begin{array}{ccc}
\mathbb{Z} \cong \mathbf{K}_{\mathbb{Z}/4}^{-1}(X_1, [\gamma] \cup \{\Gamma\}) & \xrightarrow{\times 2} & \mathbf{K}_{\mathbb{Z}/4}^{-1}(X_1, [\gamma]) \cong \mathbb{Z} \\
\downarrow & & \downarrow \\
\mathbb{Z}^2 \cong \mathbf{K}_{\mathbb{Z}/4}^{-1}(X_1, X_0) & \longrightarrow & \mathbf{K}_{\mathbb{Z}/4}^{-1}(X_1)
\end{array}
\qquad (3.21)
$$

implies that the image of $\mathbf{K}_{\mathbb{Z}/4}^{-1}(X_1, X_0)$ does not include a generator of the torsion free part of $\mathbf{K}_{\mathbb{Z}/4}^{-1}(X_1)$. From diagram (3.19) we have that the image of the first differential does not equal the image of the connecting homomorphism ∂^1. We conclude that $\mathbf{K}_{\mathbb{Z}/4}^0(X_2)$ is torsion free and the third page of the spectral sequence becomes:

$$
\begin{array}{lll}
E_3^{0,0} = \mathbb{Z}^4 & E_3^{1,0} = 0 & E_3^{2,0} = \mathbb{Z} \\
E_3^{0,-1} = \mathbb{Z}/2 & E_3^{1,-1} = 0 & E_3^{2,-1} = 0 \\
E_3^{0,-2} = (\mathbb{Z}/2)^2 \oplus \mathbb{Z} & E_3^{1,-2} = 0 & E_3^{2,-2} = 0 \\
E_3^{0,-3} = 0 & E_3^{1,-3} = 0 & E_3^{2,-3} = 0
\end{array}
\qquad (3.22)
$$

collapsing on the E_3-page, c.i. $E_3 = E_\infty$. We now look at the filtration of the magnetic K-theory groups and solve the extension problems presented in Remark 2.318

$$0 \longrightarrow E_\infty^{2,-n-2} \longrightarrow F^1 \mathbf{K}_{\mathbb{Z}/4}^{-n}(T^2) \longrightarrow E_\infty^{1,-n-1} \longrightarrow 0 \tag{3.23}$$

$$0 \longrightarrow F^1 \mathbf{K}_{\mathbb{Z}/4}^{-n}(T^2) \longrightarrow \mathbf{K}_{\mathbb{Z}/4}^{-n}(T^2) \longrightarrow E_\infty^{0,-n} \longrightarrow 0 \tag{3.24}$$

to recover the $\mathbf{K}_G^*$-groups:

- $n = 0$: $F^1 \mathbf{K}_{\mathbb{Z}/4}^0(T^2) = 0$ and $\mathbf{K}_{\mathbb{Z}/4}^0(T^2) = \mathbb{Z}^4$.
- $n = -1$: $F^1 \mathbf{K}_{\mathbb{Z}/4}^{-1}(T^2) = 0$ and $\mathbf{K}_{\mathbb{Z}/4}^{-1}(T^2) = \mathbb{Z}/2$.
- $n = -3$: $F^1 \mathbf{K}_{\mathbb{Z}/4}^{-3}(T^2) = 0$ and $\mathbf{K}_{\mathbb{Z}/4}^{-3}(T^2) = 0$.

The only extension that needs attention is the one for $n = -2$. Here $F^1 \mathbf{K}_{\mathbb{Z}/4}^{-2}(T^2) = \mathbb{Z}$ and the short exact sequence becomes:

$$0 \longrightarrow \mathbb{Z} \longrightarrow \mathbf{K}_{\mathbb{Z}/4}^{-2}(T^2) \longrightarrow \mathbb{Z} \oplus (\mathbb{Z}/2)^2 \longrightarrow 0. \tag{3.25}$$

One copy of $\mathbb{Z}/2$ splits from the short exact sequence because the composition of the $\mathbb{Z}/4$ equivariant maps $\{\Gamma\} \hookrightarrow T^2 \to *$ implies that the map $\mathbb{Z}/2 \cong \mathbf{K}_{\mathbb{Z}/4}^{-2}(*) \hookrightarrow \mathbf{K}_{\mathbb{Z}/4}^{-2}(T^2)$ is injective. We claim that the other copy of $\mathbb{Z}/2$ does not split.

Consider the quotient $T^2 / \mathbb{Z}/2$ where $\mathbb{Z}/2 = \ker(\mathbb{Z}/4 \to \mathbb{Z}/2)$ and note that it is homeomorphic to $S^{2,1}$, the 2-sphere with the involution given by a rotation of 180 degrees. The quotient map $q : T^2 \longrightarrow S^{2,1}$ induces the following commutative diagrams:

$$\tag{3.26}$$

The first row of diagram (3.26) is exact because the real K-theory of the pair $(S^{2,1}, q(X))$ is equivalent to the one of the pair $(B^{2,0}, \partial B^{2,0})$ and $K\mathbb{R}^{-1}$ $(B^{2,0}, \partial B^{2,0}) = 0$. Since the vertical map in the bottom of the right-hand side of diagram (3.26) is an isomorphism, we see that the sequence in the bottom row of of diagram (3.26) does not split. We therefore conclude that the sequence in diagram (3.25) does not split; thus implying that

$$\mathbf{K}_{\mathbb{Z}/4}^{-2}(T^2) = (\mathbb{Z} \oplus \mathbb{Z}/2) \oplus \mathbb{Z}. \tag{3.27}$$

We conclude that the magnetic $\mathbb{Z}/4$-equivariant K-theory groups of the 2-dimensional torus are given as follows.

Proposition 3.1 *The magnetic $\mathbb{Z}/4$-equivariant K-theory groups of the 2-dimensional torus T^2 are:*

$$
\begin{array}{c|cccc}
n & 0 & -1 & -2 & -3 \\
\hline
\mathbf{K}^n_{\mathbb{Z}/4}(T^2) & \mathbb{Z}^4 & \mathbb{Z}/2 & (\mathbb{Z} \oplus \mathbb{Z}/2) \oplus \mathbb{Z} & 0
\end{array}
\tag{3.28}
$$

Here we want to emphasize that the right hand side copy of $\mathbb{Z}$ in $\mathbf{K}^{-2}_{\mathbb{Z}/4}(T^2)$ is the only invariant that depends on the 2-cell in the torus. This invariant is therefore called *bulk* invariant in condensed matter physics.

3.3 2-Dimensional Torus with $C_4\mathbb{T}$ Symmetry and Spin-Orbit Interaction

Now we are interested in calculating the magnetic equivariant K-theory groups of the group generated by $C_4\mathbb{T}$ acting on the 2-dimensional torus in the framework of spin orbit interaction. Following Definition 3.2 we are interested in the twisted magnetic K-theory groups $\widetilde{{}^{\mathbb{Z}/4}}{}^{soc}\mathbf{K}^*_{\mathbb{Z}/4}(T^2)$.

In the presence of spin orbit interaction, the rotation C_4 lifts to an operator $\widetilde{C_4}$ with the property that $(\widetilde{C_4})^4 = -1$, and $\mathbb{T}^2 = -1$. Therefore the composition $\widetilde{C_4}\mathbb{T}$ is of order 8 with $(\widetilde{C_4}\mathbb{T})^4 = -1$. The SOC group $\widetilde{\mathbb{Z}/4}^{soc}$ of $\mathbb{Z}/4 = \langle C_4\mathbb{T}\rangle$ is therefore the cyclic group $\mathbb{Z}/8 \cong \langle \widetilde{C_4}\mathbb{T}\rangle$ and we will calculate this twisted magnetic equivariant K-theory group

$$
\widetilde{{}^{\mathbb{Z}/4}}{}^{soc}\mathbf{K}^*_{\mathbb{Z}/4}(T^2) = {}^{\mathbb{Z}/8}\mathbf{K}^*_{\mathbb{Z}/4}(T^2).
\tag{3.29}
$$

Using the cell decomposition of Eqs. (3.7), (3.8) and (3.9) we get for the first page of the Atiyah-Hirzebruch spectral sequence the groups:

$$
E_1^{0,*} \cong {}^{\mathbb{Z}/8}\mathbf{K}^*_{\mathbb{Z}/4}(\{\Gamma\}) \oplus {}^{\mathbb{Z}/4}KU^*_{\mathbb{Z}/2}(\{A\}) \oplus {}^{\mathbb{Z}/8}\mathbf{K}^*_{\mathbb{Z}/4}(\{X\}),
\tag{3.30}
$$

$$
E_1^{1,*} \cong {}^{\mathbb{Z}/2}KU^*(\{\sigma\}) \oplus {}^{\mathbb{Z}/2}KU^*(\{\gamma\}),
\tag{3.31}
$$

$$
E_1^{2,*} \cong {}^{\mathbb{Z}/2}KU^*(\{\Omega\}).
\tag{3.32}
$$

For the 0-cells Γ and X we know that the only non-trivial $\mathbb{Z}/8$-twisted magnetic irreducible representation of $\mathbb{Z}/4$ is generated by the magnetic matrix $\left(\begin{smallmatrix} 0 & i \\ 1 & 0 \end{smallmatrix}\right)\mathbb{K}$, and since it squares to $\left(\begin{smallmatrix} i & 0 \\ 0 & -i \end{smallmatrix}\right)$ we see that this representation is of complex type. Abusing of notation, we obtain:

$$
{}^{\mathbb{Z}/8}\mathbf{K}^t_{\mathbb{Z}/4}(\{\Gamma\}) = {}^{\mathbb{Z}/8}\mathbf{K}^t_{\mathbb{Z}/4}(\{X\}) =
\begin{cases}
{}^{\mathbb{Z}/8}\mathbf{R}(\mathbb{Z}/4) = \mathbb{Z}\langle R = \left(\begin{smallmatrix} 0 & i \\ 1 & 0 \end{smallmatrix}\right)\mathbb{K}\rangle & \text{if } t \equiv 0 \bmod 2 \\
0 & \text{if } t \equiv 1 \bmod 2.
\end{cases}
\tag{3.33}
$$

The 0-cells $A \sqcup A'$ have as stabilizer the non magnetic group $\mathbb{Z}/4$ which is an extension of $\mathbb{Z}/2$ by the core group $\mathbb{Z}/2$. The irreducible complex representations of $\mathbb{Z}/4$ where $\mathbb{Z}/2$ acts by multiplication by -1 are the ones generated by i and $-i$. If we denote the representations by their generators we get:

$$^{\mathbb{Z}/4}KU^t_{\mathbb{Z}/2}(\{A\}) = \begin{cases} ^{\mathbb{Z}/4}R(\mathbb{Z}/2)) = \mathbb{Z}\langle i, -i \rangle & \text{if } t \equiv 0 \bmod 2 \\ 0 & \text{if } t \equiv 1 \bmod 2. \end{cases} \tag{3.34}$$

For the 1-cells and 2-cells where the stabilizer is trivial we simply obtain the complex K-theory group where $\mathbb{Z}/2$ acts on the fibers via the sign representation:

$$^{\mathbb{Z}/2}KU^*(\{\sigma\}) = {}^{\mathbb{Z}/2}KU^*(\{\gamma\}) = {}^{\mathbb{Z}/2}KU^*(\{\Omega\}) = \begin{cases} \mathbb{Z}\langle \text{sgn} \rangle & \text{if } t \equiv 0 \bmod 2 \\ 0 & \text{if } t \equiv 1 \bmod 2. \end{cases} \tag{3.35}$$

Since the twisted magnetic equivariant K-theory group $^{\mathbb{Z}/8}\mathbf{K}^*_{\mathbb{Z}/4}(T^2)$ is a subgroup of the equivariant magnetic K-theory $\mathbf{K}^*_{\mathbb{Z}/8}(T^2)$, then we know from Proposition 2.6 that the K-theory groups are 4-periodic. We will only calculate $^{\mathbb{Z}/8}\mathbf{K}^*_{\mathbb{Z}/4}(T^2)$ for $* = 0, -1, -2, -3$.

The first page of the spectral sequence then becomes:

$$\begin{aligned} E_1^{0,0} = \mathbb{Z} \oplus \mathbb{Z}^2 \oplus \mathbb{Z} &\xrightarrow{d_1^{0,0}} E_1^{1,0} = \mathbb{Z} \oplus \mathbb{Z} \xrightarrow{d_1^{1,0}} E_1^{2,0} = \mathbb{Z} \\ E_1^{0,-1} = 0 &\xrightarrow{d_1^{0,-1}} E_1^{1,-1} = 0 \xrightarrow{d_1^{1,-1}} E_1^{2,-1} = 0 \\ E_1^{0,-2} = \mathbb{Z} \oplus \mathbb{Z}^2 \oplus \mathbb{Z} &\xrightarrow{d_1^{0,-2}} E_1^{1,-2} = \mathbb{Z} \oplus \mathbb{Z} \xrightarrow{d_1^{1,-2}} E_1^{2,-2} = \mathbb{Z} \\ E_1^{0,-3} = 0 &\xrightarrow{d_1^{0,-3}} E_1^{1,-3} = 0 \xrightarrow{d_1^{1,-3}} E_1^{2,-3} = 0 \end{aligned} \tag{3.36}$$

The differential $d_1^{0,0}$ keeps the rank of the dual of the restrictions. Hence we have:

$$d_1^{0,0} = \begin{array}{|c|cc|c|c|} \hline \Gamma & \multicolumn{2}{c|}{A} & X & \\ R & i & -i & R & \\ \hline 0 & -1 & -1 & 2 & \gamma \\ -2 & 1 & 1 & 0 & \sigma \\ \hline \end{array} \tag{3.37}$$

The next differential satisfies $d_1^{0,0} = 0$ because the boundary of Ω is zero in terms of γ and σ.

The first differential on the row -2 becomes:

$$d_1^{0,-2} = \begin{array}{|c|cc|c|c|} \hline \Gamma & \multicolumn{2}{c|}{A \sqcup A'} & X & \\ R' & i & -i & R' & \\ \hline 0 & -1 & -1 & 0 & \gamma \\ 0 & 1 & 1 & 0 & \sigma \\ \hline \end{array} \tag{3.38}$$

where the zeroes from the first and the third columns come from the fact that the forgetful homomorphism ${}^{\mathbb{Z}/8}\mathbf{K}^{2}_{\mathbb{Z}/4}(*) \to KU^{2}(*)$ is trivial. The reason for this is the following. A $\mathbb{Z}/8$-twisted magnetic $\mathbb{Z}/4$-equivariant vector bundle over S^2 (here the action of $\mathbb{Z}/4$ is trivial) can be split into the eigenbundles of i and $-i$ respectively. The action of the generator $\left(\begin{smallmatrix} 0 & i \\ 1 & 0 \end{smallmatrix}\right)\mathbb{K}$ maps one to the other, but flips the complex structure. Hence the first Chern class of the underlying complex bundle is the sum of the Chern classes of the two, but they have opposite Chern class. Therefore the Chern class of the underlying complex vector bundle is zero and therefore the forgetful map is trivial.

The remaining differential has the same structure as the one presented without spin orbit interaction in Eq. (3.17):

$$d_1^{1,-2} = \begin{vmatrix} \sigma & \gamma & \\ 2 & 2 & \Omega \end{vmatrix} \tag{3.39}$$

The second page of the spectral sequence is therefore:

$$
\begin{array}{lll}
E_2^{0,0} = \mathbb{Z}^2 & E_2^{1,0} = \mathbb{Z}/2 & E_2^{2,0} = \mathbb{Z} \\
E_2^{0,-1} = 0 & E_2^{1,-1} = 0 & E_2^{2,-1} = 0 \\
E_2^{0,-2} = \mathbb{Z}^3 & E_2^{1,-2} = 0 & E_2^{2,-2} = \mathbb{Z}/2 \\
E_2^{0,-3} = 0 & E_2^{1,-3} = 0 & E_2^{2,-3} = 0
\end{array}
\tag{3.40}
$$

All the differentials $d_2^{n,t}$ are zero and the spectral sequence collapses at the second page. In this case there are no extension problems for the K-theory groups as presented in Rem 2.318 and therefore we obtain the desired calculation.

Proposition 3.2 *The $\mathbb{Z}/8$-twisted magnetic $\mathbb{Z}/4$-equivariant K-theory groups of the 2-dimensional torus T^2 are:*

$$
\begin{array}{c|cccc}
n & 0 & -1 & -2 & -3 \\
\hline
{}^{\mathbb{Z}/8}\mathbf{K}^n_{\mathbb{Z}/4}(T^2) & \mathbb{Z}^2 \oplus \mathbb{Z}/2 & 0 & \mathbb{Z}^3 \oplus \mathbb{Z} & \mathbb{Z}/2
\end{array}
\tag{3.41}
$$

Here we emphasize that both the $\mathbb{Z}/2$ invariant in ${}^{\mathbb{Z}/8}\mathbf{K}^0_{\mathbb{Z}/4}(T^2)$ as well as the right hand side copy of $\mathbb{Z}$ in ${}^{\mathbb{Z}/8}\mathbf{K}^{-2}_{\mathbb{Z}/4}(T^2)$ are invariants that depend on the 2-cell in the torus. These invariants are *bulk* invariants of the system.

Remark 3.1 The bulk $\mathbb{Z}/2$ invariant in ${}^{\mathbb{Z}/8}\mathbf{K}^0_{\mathbb{Z}/4}(T^2)$ was shown by González-Hernández and the first two authors in [13] to be the indicator for topological insulators in the presence of the altermagnetic symmetry $C_4\mathbb{T}$ [10–12]. Gapped Hamiltonians preserving the symmetry $C_4\mathbb{T}$ which represent the non-trivial value of $\mathbb{Z}/2$ are insulating states which are topological non-trivial. Hence its classification as topological insulator.

Remark 3.2 Both calculations presented in Propositions 3.1 and 3.2 agree with the calculations carried out by Shiozaki and Ono in [14].

References

1. Bellissard, J., van Elst, A., Schulz-Baldes, H.: The noncommutative geometry of the quantum Hall effect. J. Math. Phys. **35**(10), 5373–5451 (1994). Topology and physics
2. Bellissard, J.: K-theory of C^*-algebras in solid state physics. In: Statistical mechanics and field theory: mathematical aspects (Groningen, 1985), volume 257 of Lecture Notes in Physics, pp. 99–156. Springer, Berlin (1986)
3. Kitaev, A.: Periodic table for topological insulators and superconductors. In: AIP Conference Proceedings, vol. 1134, pp. 22–30. American Institute of Physics (2009)
4. Altland, A., Zirnbauer, M.R.: Nonstandard symmetry classes in mesoscopic normal-superconducting hybrid structures. Phys. Rev. B **55**, 1142–1161 (1997)
5. Cornfeld, E., Carmeli, S.: Tenfold topology of crystals: Unified classification of crystalline topological insulators and superconductors. Phys. Rev. Res. **3**, 013052 (2021)
6. Freed, D.S., Moore, G.W.: Twisted equivariant matter. Ann. Henri Poincaré **14**(8), 1927–2023 (2013)
7. Connes, A.: Géométrie Non Commutative. InterEditions, Paris (1990)
8. Prodan, E., Schulz-Baldes, H.: Bulk and boundary invariants for complex topological insulators. Mathematical Physics Studies. Springer, [Cham] (2016). From K-theory to physics
9. Gomi, K.: Freed-Moore K-theory. Commun. Anal. Geom. **31**(4), 979–1067 (2023)
10. Šmejkal, L., Sinova, J., Jungwirth, T.: Beyond conventional ferromagnetism and antiferromagnetism: a phase with nonrelativistic spin and crystal rotation symmetry. Phys. Rev. X **12**, 031042 (2022)
11. González-Hernández, R., Šmejkal, L., Výborný, K., Yahagi, Y., Sinova, J., Jungwirth, T.š., Železný, J.: Efficient electrical spin splitter based on nonrelativistic collinear antiferromagnetism. Phys. Rev. Lett. **126**:127701 (2021)
12. Mazin, I.: Editorial: altermagnetism–a new punch line of fundamental magnetism. Phys. Rev. X **12**, 040002 (2022)
13. González-Hernández, R., Serrano, H., Uribe, B.: Spin chern number in altermagnets. Phys. Rev. B **111**, 085127 (2025)
14. Shiozaki, K., Ono, S.: Atiyah-Hirzebruch spectral sequence for topological insulators and superconductors: E2 pages for 1651 magnetic space groups, Apr. 2023
15. Bárcenas, J.E., Joachim, M., Uribe, B.: Universal twist in equivariant K-theory for proper and discrete actions. Proc. Lond. Math. Soc. (3), **108**(5):1313–1350 (2014)

Conclusions, and Further Remarks

Conclusions

We have presented several of the most important properties of the K-theory of magnetic equivariant vector bundles. We have done so emphasizing simplicity in arguments, proofs and notation. We have carried out explicit calculations of these K-theory groups for important symmetries in condensed matter physics, and we have obtained important topological invariants associated to these symmetries.

We have complemented and enhanced the works of Freed and Moore [1] and Gomi [2] in the magnetic equivariant K-theory, and in doing so, we have put forward a better suited name for these K-theory groups.

We believe that with the properties of the magnetic equivariant K-theory presented in this work, the calculation of these K-theory groups for explicit space magnetic groups becomes accessible for both mathematicians and condensed matter physicists. Important here is to notice that only symmorphic magnetic crystal symmetries can be addressed with the results of this work. Non symmorphic magnetic crystal symmetries require the study of twisted magnetic equivariant K-theory where the twists incorporate both information of the point group and of the torus.

One of the motivations to address and showcase the properties of the magnetic equivariant K-theory was the necessity of putting on safe grounds the various calculations of magnetic equivariant K-theory groups that were done by the three authors in previous works [3–5]. These works dealt with the topological invariants associated to a prescribed symmetry (in our works these were $C_2\mathbb{T}$ and $C_4\mathbb{T}$) and the calculations of the magnetic equivariant K-theory were done assuming they possessed the properties we showcase in the previous chapters. This work fills this foundational gap.

We hope that our presentation of the properties of the magnetic equivariant K-theory will allow more people to determine topological invariants of magnetic materials. We are sure that further interesting topological features of magnetic crystals are going to be discovered.

B. Uribe Jongbloed et al., *Magnetic Equivariant K-Theory*,
SpringerBriefs in Mathematics, https://doi.org/10.1007/978-3-032-05914-7

Further Remarks

The present work presents a particular approach to construct topological invariants that detect non-trivial quantum systems of free fermions in a periodic potential. However, there are several other mathematical frameworks which also yield topological invariants. We invite the interested reader to consult the following references for an overview of these alternative approaches. We clarify that this list is by no means intended to be exhaustive.

- Non-commutative geometry. One can model the interactions of particles in a certain region of the space and their symmetries using C^*-algebras and then take their K-theory or K-homology to produce invariants that capture topological features of the system. The key idea is that, in certain quantum setting, classical spaces are no longer appropriate and must by replaced by non-commutative algebras of observables. This path was brilliantly initiated by Bellissard and his collaborators [6, 7] using the framework of non-commutative geometry á la Connes [8]. The book of Prodan and Schulz-Baldes [9] offers a great introduction to the ideas, has numerous references and moreover tackles the important question of relating the bulk and the edge invariants.
- Chern-Simons theory. This topological quantum field theory associates invariants to smooth manifolds by integrating certain characteristic classes from a gauge connection. In condensed matter, Chern-Simons theory often describes low energy effective actions whose quantization leads to topological invariants distinguishing different phases, such as in the quantum spin Hall effect. See for example [10–12] or the comprehensive work of Freed on the subject [13–15].
- Bordism. In this framework, topological phases of matter are classified using ideas from cobordism theory. The basic idea is that two physical systems are equivalent if they can be connected by a continuous deformation in a higher dimensional system. This leads to a classification of phases in terms of bordism groups with additional structure as spin, spin_c, orientation or symmetry. See for example [16–19] and the references therein.

The magnetic equivariant K-theory groups presented in this work are one example of what is known as an equivariant cohomology theory [20]. Equivariant cohomology theories can be obtained from the associated equivariant spectra and this approach has been the subject of intense research by homotopy theorists. We recommend the following books for a comprehensive account on this matter [21–24]. The twisted magnetic equivariant K-theory groups can be understood from the point of view of a parametrized equivariant spectra [25]. The second author with collaborators used this structure to study twisted equivariant K-theory [26, 27]. A similar perspective from the point of parametrized equivariant homotopy theory should be possible be constructed for the twisted magnetic equivariant K-theory. We leave this for the interested reader.

The analytical counterpart of K-theory is also a well-developed, amazingly rich and very important branch of mathematics where analysis, algebra, geometry and topology meet. K-homology defined via Fredholm-Hilbert modules [28], its generalization to KK-theory by Kasparov [29, 30] incorporating both K-theory and K-homology, together with its equivariant version [31], opened new paths in the understanding of C^*-algebras, their classification and their invariants, among others [32]. The recent book of Lück [33] has an up-to-date account of the state of the isomorphism conjectures together with a detailed account of some of the most important results in the area.

References

1. Freed, D.S., Moore, G.W.: Twisted equivariant matter. Ann. Henri Poincaré **14**(8), 1927–2023 (2013)
2. Gomi, K.: Freed-Moore K-theory. Commun. Anal. Geom. **31**(4), 979–1067 (2023)
3. González-Hernández, R., Pinilla, C., Uribe, B.: Axion insulators protected by $C_2\mathbb{T}$ symmetry, their K-theory invariants, and material realizations. Phys. Rev. B **106**, 195144 (Nov2022)
4. Serrano, H., Uribe, B., Xicoténcatl, M.A.: Rational magnetic equivariant K-theory. Rev. Acad. Colombiana Cienc. Exact. Fís. Natur. **49**(190), 183–197 (2025)
5. González-Hernández, R., Serrano, H., Uribe, B.: Spin chern number in altermagnets. Phys. Rev. B **111**, 085127 (2025)
6. Bellissard, J.: K-theory of C^*-algebras in solid state physics. In: Statistical mechanics and field theory: mathematical aspects (Groningen, 1985), volume 257 of Lecture Notes in Physics, pp. 99–156. Springer, Berlin (1986)
7. Bellissard, J., van Elst, A., Schulz-Baldes, H.: The noncommutative geometry of the quantum Hall effect. J. Math. Phys. **35**(10), 5373–5451 (1994). Topology and physics
8. Connes, A.: Géométrie non Commutative. InterEditions, Paris (1990)
9. Prodan, E., Schulz-Baldes, H.: Bulk and boundary invariants for complex topological insulators. Mathematical Physics Studies. Springer, [Cham] (2016). From K-theory to physics
10. De Nittis, G., Gomi, K.: Classification of "quaternionic" bloch-bundles: topological quantum systems of type AII. Commun. Math. Phys. **339**(1), 1–55 (2015)
11. Dunk, S., Szabo, R.J.: Topological insulators and the kane–mele invariant: obstruction and localization theory. Rev. Math. Phys. **32**(06), 2050017 (2020)
12. Andrei Bernevig, B., Zhang, S.-C.: Quantum spin hall effect. Phys. Rev. Lett. **96**, 106802 (2006)
13. Freed, D.S.: Classical Chern-Simons theory. I. Adv. Math. **113**(2), 237–303 (1995)
14. Freed, D.S.: Classical Chern-Simons theory. II. Special issue for S. S. Chern (2002)
15. Freed, D.S.: Remarks on Chern-Simons theory. Bull. Am. Math. Soc. (N.S.) **46**(2), 221–254 (2009)
16. Wan, Z., Wang, J.: Higher anomalies, higher symmetries, and cobordisms I: classification of higher-symmetry-protected topological states and their boundary fermionic/bosonic anomalies via a generalized cobordism theory. Ann. Math. Sci. Appl. **4**(2), 107–311 (2019)
17. Freed, D.S., Hopkins, M.J.: Reflection positivity and invertible topological phases. Geom. Topol. **25**(3), 1165–1330 (2021)
18. Schommer-Pries, C.: Invertible topological field theories. J. Topol. **17**(2), Paper No. e12335, 64 (2024)
19. Stehouwer, L.: Interacting SPT phases are not Morita invariant. Lett. Math. Phys. **112**(3), Paper No. 64, 25 (2022)
20. May, J.P.: Equivariant homotopy and cohomology theory. In: Symposium on Algebraic Topology in honor of José Adem (Oaxtepec, 1981), volume 12 of Contemporary Mathematics, pp. 209–217. American Mathematical Society, Providence, RI (1982)

21. Lewis, Jr. L.G., May, J.P., Steinberger, M., McClure, J.E.: Equivariant stable homotopy theory, volume 1213 of Lecture Notes in Mathematics. Springer, Berlin (1986). With contributions by J.E. McClure
22. Greenlees, J.P.C., May, J.P.: Equivariant stable homotopy theory. In: Handbook of Algebraic Topology, pp. 277–323. North-Holland, Amsterdam (1995)
23. May, J.P.: Equivariant homotopy and cohomology theory, volume 91 of CBMS Regional Conference Series in Mathematics. Conference Board of the Mathematical Sciences, Washington, DC; by the American Mathematical Society, Providence, RI (1996). With contributions by M. Cole, G. Comezaña, S. Costenoble, A.D. Elmendorf, J.P.C. Greenlees, L.G. Lewis, Jr., R.J. Piacenza, G. Triantafillou, and S. Waner
24. Degrijse, D., Hausmann, M., Luck, W., Patchkoria, I., Schwede, S.: Proper equivariant stable homotopy theory. Mem. Am. Math. Soc. 288(1432), vi+142 (2023)
25. May, J.P., Sigurdsson, J.: Parametrized homotopy theory. Mathematical Surveys and Monographs, vol. 132. American Mathematical Society, Providence, RI (2006)
26. Bárcenas, N., Espinoza, J., Joachim, M., Uribe, B.: Universal twist in equivariant K-theory for proper and discrete actions. Proc. Lond. Math. Soc. (3) 108(5), 1313–1350 (2014)
27. Bárcenas, N., Espinoza, J., Uribe, B., Velásquez, M.: Segal's spectral sequence in twisted equivariant K-theory for proper and discrete actions. Proc. Edinb. Math. Soc. (2), 61(1), 121–150 (2018)
28. Higson, N., Roe, J.: Analytic K-homology. Oxford Mathematical Monographs. Oxford University Press, Oxford (2000). Oxford Science Publications
29. Kasparov, G.G.: The operator K-functor and extensions of C^*-algebras. Izv. Akad. Nauk SSSR Ser. Mat. 44(3):571–636, 719 (1980)
30. Skandalis, G.: Kasparov's bivariant K-theory and applications. Exposition. Math. 9(3), 193–250 (1991)
31. Kasparov, G.G.: Equivariant KK-theory and the Novikov conjecture. Invent. Math. 91(1), 147–201 (1988)
32. Blackadar, B.: Operator algebras, volume 122 of Encyclopaedia of Mathematical Sciences. Springer, Berlin: Theory of C^*-algebras and von Neumann algebras. Operator Algebras and Non-commutative Geometry, III (2006)
33. Lück, W.: Isomorphism Conjectures in K- and L-Theory, volume 122 of Ergebnisse der Mathematik und ihrer Grenzgebiete. 3. Folge/A Series of Modern Surveys in Mathematics. Springer, Berlin (2025)